PERSPECTIVES IN GEOMORPHOLOGY

VOLUME TWO

By the Same Author

- Climate Change: Bio-Diversity and Green Economy with *S. Padmaja* and *Ganesh Sharma*
- Indian Geomorphology
- Mathematical Modelling in Geographical Information System Global Positioning System and Digital Cartography with *D.R. Ram, Rama Prasad* and *P.R. Binda*
- Modeling in Resource Management and Environment Through Geomatics with *P.R. Binda*
- Ranthambhore Sanctuary: Dilemma of Eco-Development
- Ravine Erosion in India
- Sustainable Development: Issues and Case Studies with *Sri K. Chattopadhyay*
- Tropical Geomorphology: A Morphogenetic Study of Rajasthan

About the Editor

Prof. H. S. Sharma (*b.* 1945) holds post-graduate and doctoral degrees in Geography from the University of Saugar, Madhya Pradesh. Prof. Sharma is a former Professor and Head, Department of Geography, University of Rajasthan, Jaipur. He is also a former Director of the Indira Gandhi Centre for Human Ecology, Environment and Population Studies (HEEPS), Academic Staff College and Dean, Faculty of Science, University of Rajasthan.

Prof. Sharma has over 100 research papers in National and International journals like *Nature* and 16 books to his credit. He has supervised more than three dozen Ph.D.s. Prof. Sharma has travelled widely across U.K., U.S.A., Germany, Holland, Canada, Norway, Seychelles and Malaysia to attend and chair technical sessions in International Conferences and Exchange Programmes.

Prof. Sharma has successfully conducted several research projects sponsored by the U.G.C., Ministry of Environment and Forests, DRDO, New Delhi, NORAD, Norway and ISRO (SAC) Ahmedabad.

Prof. Sharma has been President of IIG, IGI and NAGI. He is also a President of the Commonwealth Human Ecology Council (India Chapter). Recently, he has been awarded Senior Fellowship by Indian Council of Social Science Research, New Delhi to work on Climate Change and its Impact on Land and Water Resources of Rajasthan.

Concept's International Series in Geography No. 2

PERSPECTIVES IN GEOMORPHOLOGY

VOLUME TWO

QUANTITATIVE FLUVIAL GEOMORPHOLOGY

Edited by

H.S. Sharma

CONCEPT PUBLISHING COMPANY PVT. LTD.
NEW DELHI-110059

ISBN-13: 978-81-7022-123-4 (Set)

Edition 2016

Published and Printed by

Concept Since 1974

Concept Publishing Company Pvt. Ltd.
A/15-16, Commercial Block, Mohan Garden
New Delhi-110059 (India)
T: 25351460, 25351794, *F*: 091-11-25357109
E: publishing@conceptpub.com, W: www.conceptpub.com
Editorial Office: H-13, Bali Nagar, New Delhi-110 015, India.

Cataloging in Publication Data--Courtesy: D.K. Agencies (P) Ltd. <docinfo@dkagencies.com>

Perspectives in geomorphology / edited by H.S. Sharma.
volumes cm. -- (Concept's international series in geography ; no. 2)
Reprint. Originally published: 1981-1982.
Includes bibliographical references and index.
Contents: volume 1. Recent trends -- volume 2. Quantitative fluvial geomorphology -- volume 3. Applied geomorphology volume 4. Essays on Indian geomorphology.
ISBN 9788170221234 (set)

1. Geomorphology. 2. Geomorphology--India. I. Sharma, H. S. (Hari Shanker), 1945- editor. II. Series: Concept's international series in geography ; no. 2.

GB401.5.P47 2016 DDC 551.4 23

PUBLISHER'S NOTE

THIS IS the second set of volumes in our International Geography Series. It follows close upon the heels of the five-volume set on *Perspectives in Agricultural Geography* edited by Dr. Noor Mohammad. We take this opportunity at the very outset to acknowledge the co-operation we have received from the Editor Dr. Noor Mohammad and the numerous contributors from all over the world who have made Series One possible. The wide acclaim and the heartening response from our readers—researchers and students in the field of agricultural geography—have encouraged us to plan the publication of this second set entitled *Perspectives in Geomorphology* edited by Dr. Hari Shanker Sharma. Here again we would like to thank the Editor and the contributors from various countries of the world who have given a truly international perspective to these volumes. We are confident that this series will also be equally well-received all over the world.

The Concept Series in Geography is now an on-going one and we are very happy to announce that plans are afoot to make available a third set of volumes in the very contemporary subject of Urban Geography. This series is expected to run into fifty volumes and the publication programme, hopefully, will commence early in 1982. The response to our brochure has been very encouraging and more than 150 articles have already been received including several from overseas contributors.

With the publication of the initial volumes of this new series we hope to announce the subject of the fourth series so that both students and prospective contributors will receive information in advance.

In the next few years we hope in this manner, to provide geographers with a complete compendium of the latest advances in major areas in geography. Here, we might emphasize again, that the benefits that the students in this country as well as in other developing countries will obtain will be immeasurable.

Finally, we crave once again, the indulgence of our readers for any shortcomings in these volumes. We shall learn with experience.

PREFACE

IN THE last three decades there has been considerable progress in all branches of science. The science of geomorphology has also achieved a great deal during this time, and geographers, geologists, hydrologists, pedologists and specialists from other fields who have contributed to the study of geomorphology, have become aware of the comparative suddenness of the change within the context of land form studies. As a matter of fact, it is becoming virtually impossible for a single scholar to command all aspects of geomorphology. However, there has been increasing specialization at all levels on the one hand and a need for synthesising the results of these researches, to benefit the scholars working in the field of geo-science.

The present project which has been sponsored by the Concept Publishing Company, New Delhi is an humble attempt to disseminate the knowledge of concepts, theories, models and recent trends in geomorphology with reference to the present conceptual revolution in this field. With this ambition in view original research papers dealing with modern quantitative techniques were invited from geo-scientists from all over the world. The basic themes along with the tentative titles of the papers were sent to the contributors in various branches of geo-science. The contributors have responded well with original research papers.

Later on, these papers were thoroughly edited and the papers found worthy for publication were classified and arranged into four volumes.

The first volume of the series contains the papers dealing with the modern trends in geomorphology. This volume has been divided into five sections. Section one deals with the stream lines of geomorphic research, climatic and climato-genetic geomorphology, planation surfaces, granitic inselbergs and duri-crusts. Section two discusses Karst processes and land forms. Papers dealing with glacial and peri-glacial land-forms have been included in Section three. Section four contains papers dealing with the coastal processes and land forms in tropical region. Section five contains the papers dealing with remote sensing and geomorphology.

As considerable amount of work has been done in fluvial geomorphology, therefore, volume two has been devoted to the papers dealing with the various aspects of quantitative fluvial geomorphology.

Volume three contains papers on applied geomorphology. Papers dealing with land classification, geomorphological mapping, terrain evaluation, and application of photo-interpretation and remote sensing in geomorphological mapping have been included in this volume.

Geo-scientists in India have also become aware and realised the significance of land form studies in their researches. As a matter of fact considerable amount of work has been done in various aspects of geomorphology. The fourth volume of the series contains papers on Indian geomorphology. It is earnestly hoped that these volumes will serve the purpose of enlightening the readers regarding the kind of research which is being taken up by modern geomorphologists today. As the editor of these volumes and, in addition a contributor, I count myself highly fortunate in my fellow contributors. All those whose papers are included in these volumes have prepared and supplied their material in the face of manifest difficulties—mainly the difficulties of time which pressed hard upon every busy academician. Most of my fellow contributors have patiently dealt with manuscripts and with queries relating to their papers. Those who sent their materials early have been tolerant of the subsequent delay, as late comers have tolerantly borne pressure from myself to complete their papers.

It would be ungenerous to conclude this introduction without acknowledging the guidance, encouragement and help of a number of senior scientists.

I am indebted to Professor Moonis Raza for his suggestions in making the outline of the project during my visit to the Centre for Regional Development, Jawaharlal Nehru University, New Delhi from January to April, 1980. I am highly thankful to Prof. Vaidyanadhan of Waltair University and Prof. Pramila

Kumar of Saugar University for their suggestions and encouragement. I am also thankful to Shri P.S. Dhinwa and Shri Ram Avatar Sharma for their assistance throughout the preparation of these volumes. Staff members of the Geography Department, University of Rajasthan, Jaipur, also deserve thanks for their co-operation and kindness in the preparation of this project. Special thanks are due to the publisher who has calmly suffered the repeated redistribution of authors throughout the world and who maintained effective contacts with myself. In addition I may suitably take this occasion to make a block statement of gratitude, on behalf of all the contributors, to the numerous friends and colleagues who have at various stages read part or all of the manuscripts and have supplied comments, criticisms and suggestions.

Department of Geography
Rajasthan University
Jaipur (India)

H.S. SHARMA

Independence Day
15 August 1981

CONTENTS

LIST OF CONTRIBUTORS

Athol D. Abrahams
Department of Geography, State University of New York at Buffalo, Buffalo, New York U.S.A.

Goutam Bhattacharya
Lecturer, Department of Geography, Kamaun University, Naini Tal, India

F. David Deane
Project Engineer, Anderson Nichols Consulting Engineers, Corcord, NH, USA.

V. Gardiner
Department of Geography, University of Leicester, Leicester, England

Nadira Kabir
Graduate Research Assistant, Department of Civil and Environmental Engineering, Washington State University, Pullman, U.S.A.

Alan J. Miller
Division of Mathematics and Statistics, Commonwealth Scientific and Industrial Research Corporation Lindfield, New South Wales, Australia.

John F. Orsborn
Chairman, Department of Civil and Environment Engineering, Washington State University, Pullman U.S.A.

G. Padmaja
Centre for Studies in Regional Development, Jawaharlal Nehru University, New Delhi, India.

H. S. Sharma
Career Scientist (UGC), Department of Geography, University of Rajasthan, Jaipur, India

Savindra Singh
Department of Geography, Allahabad University, Allahabad, India

Miss Soundarvalli
Department of Geography, Osmania University, Hyderabad, India

D. P. Upadhyaya
T.D. Post Graduate College, Jaunpur, India

V. K. Verma
Professor, Department of Geology, University of Delhi, Delhi, India.

LIST OF FIGURES

LIST OF TABLES

INTRODUCTION

In recent years one major conceptual change that has taken place in geomorphology has been in the quantification of fluvially eroded topography. Sophisticated quantitative techniques have been evolved to analyse the dynamic nature of the earth's form. The development of quantification is partly the result of the availability of computers, which enable large masses of data to be rapidly analysed, employing sophisticated, statistical and mathematical methods. One field of geomorphology in which quantitative analysis is earliest and has gone farthest, is that of drainage basin stream pattern analysis. The studies in this aspect have been initiated by Horton (1945) an eminent hydrologists, and Strahler (1952, 1958); and are widely published by other geomorphologists like Kesseli (1946), Miller (1953), Schumm (1956), Melton (1958), Smith (1958), Morisawa (1962), Robinson (1963), Leopold, Wolman and Miller (1964), Clark (1966), Chorley (1966) and Gardiner and Park (1978), and a number of others.

There is a danger in the quantitative study of a drainage basin, however, in that in some instances the numerical methods seem to become the goal of the study rather than nearly the means to a fuller understanding of the genesis of landforms. This is perhaps inevitable in the early stages of a major change of method, until the value of various numerical techniques can be properly assessed (C.A.M. King, 1976).

The year 1945 is essentially a major historical landmark in the growth and development of geomorphological thought. Robert E. Horton (1932, 1945) provided a new paradigm in geomorphology; pioneered the quantitative approach to drainage basin description realising the linkage between hydrological characteristics of a drainage system with their morphological characteristics. The structural functional model of Horton attracted a host of his followers who were better equipped with the tools of mathematics, (Strahler, Scheidegger 1967; Fraser Walsh, 1972), they modified and restructured Hortonian model.

Goutam Bhattacharya and V.K. Verma in their papers have reviewed the geomorphological literature of the sixties and stressed that Horton's (1945) contribution have been revolutionary because of two reasons :

1. It sowed the seed for structural functional analysis because of quantification of many geomorphic variables. This quantitative approach metamorphosed the discipline of geomorphology and brought it closer to many other disciplines making it in a real sense multidisciplinary in nature.
2. The inconsistencies in Horton's method itself were of tremendous epistemological significance. The Hortonian inconsistency fecilitated the entry of topology in geomorphology. Application of topology not only elevated the status of this discipline but also brought geomorphology closer to hydrology and water resources studies, thus, widening the social relevance of Geomorphology.

The nature of a river network must inevitably depend to some extent upon climate since the geometry of river channels is controlled by discharge and rainfall. Input to the hydrological system is modified by temperature, humidity, wind direction and the totality of such effects which we envisage as climate is manifested in terms of output from hydrological systems as runoff and evapotranspiration process. Chorley (1969) and Ward (1975) have summarised the elements of hydrological cycle.

Gardiner in his paper has stressed the significance of climate in fluvial systems. Study of this influence has proceeded from the empirical identification of largely 'black box' relationships at the basin scale in the 1950s towards investigation of smaller scale relationships in which process and theoretical modelling are involved. This has resulted from the necessity to understand the process responsible for relationships within the fluvial system. This successful integration between empirically identified relationships and process related explanations is yet incomplete and is hindered by an incomplete and spatially uneven availability of data and observations. Sufficient results are, however, available to suggest that this field may be one of considerable geomorphological value.

The drainage network and basin morphology are the measures of fluvial process operation in the basin. Studies have shown that river flows are related to the topographic and climatic characteristics of a drainage basin which control the amount, time and space distributions of stream flow. Nadira Kabir and John Orsborn in their papers have studied the quantitative relationships between the bankful channel characteristics of top width, mean depth, velocity and bed material with drainage basin characteristics through dimensional analysis at various locations in two river systems (upper Salmen and St. Mariese basins in Idaho). The results of this study demonstrate that basin, channel and flow characteristics are interrelated and can be quantified.

Exterior channel links or first order Strahler streams are sections of channel extending without intervening forks from a source at their upstream end to a fork at their downstream end (Shreve, 1966). Early investigations of exterior link-lengths in the United States, (Miller, 1953; Strahler, 1954; Schumm, 1966; Maxwell, 1960); and in a more recent study in England (Gardiner, 1973) have indicated that the frequency distributions of such lengths measured on topographic maps can be closely approximated by the log-normal probability density.

Abrahams and Miller in their papers critically made an analysis of 26 frequency distributions of exterior, source and tributary source link-lengths from 11 areas representing a variety of environments, reliefs and stages of landscape development. Their analysis suggests :

1. That the observed link length distribution of all approximate to varying degrees is single underlined probability-density which is neither the log-normal nor the gama-density; and second the log-normal density is generally superior to the gama as an empirical model for these link-lengths.

2. They further stressed to derive a simple and analytical expression for the underlying density and to identify the geomorphic mechanism giving rise to it were unsuccessful.

Though hydrologists have developed numerous models and methods for estimating low, average and flood flows at ungaged sites on streams, there are seasonal limitations on all methods :

1. Some require extensive and expensive parametric data collection systems.
2. Some require the determination of variables which at best can be considered only as approximations.
3. Only a few of the methods have more than a very limited geographic or hydrologic range of application.
4. Some of the parameters do not have stable physical relationships with the unknown flow.
5. Even fewer of the methods can be used to relate one type of stream flow (low, average or flood) to another.

Orsborn, Nadira Kabir and David Deane have developed a model taking climatically uniform provinces in Washington, Idaho and Oregon. The model will help to provide solutions to this type of problems based on factual data and physical relationships. The model described by these authors relates one characteristic flow to another (output : output) and consider the entire drainage basin as the integrator of its stream flows. Provincial correlations for average and low flows for gaged stream and various combinations of geomorphic parameters of the drainage basins above the gages were developed. To determine these flows from the ungaged watersheds, the same geomorphic parameters are determined and entered into the provincial correlations, thus, yielding the desired ungaged flows. Using the provincial correlations as developed by the authors between floods, average and low flows, flood flows in ungaged streams can then be predicted. This new model will be useful in defining water availability. They have stressed that by using the output-output method and other hydro-geomorphological methods for determining low and average flows in ungaged basins, it is possible to get a good idea of the natural flow regime of any basin, and man-made influences on that regime, with a minimum of information about the basin.

The application of morphometric techniques to geomorphic analysis of drainage basin form has provided a sound footing to this discipline. The term morphometry is used in several disciplines to mean the measurement and analysis of form characteristics; in geomorphology it is applied to numerical examination of landform, which may be more properly termed geomorphometry (Gardiner). Since 1940 number of geomorphologists have applied these techniques in their studies. Gardiner in his paper has given an up-to-date review of drainage basin morphometry and assessed the present day status and potential. The author has stressed that morphometry is potentially a most important approach to geomorphology, since it affords quantitative information on large scale fluvial landforms, which make up the vast majority of the earth surface.

Sharma and Padmaja in their paper on drainage basin morphometry have analysed various basin parameters of 20 representative basins from different geological formations and correlated with each other. The study of the Morel basin has upheld the Horton's and Strahler's models.

Savindra Singh and Upadhyaya in their papers have studied topological and geometric study of drainage network taking south-east Chota Nagpur Region. 30 basins of 6th and 5th order of different physiographic regions have been selected to study the nature of development of open and closed links of drainage network. The study concludes that the different mathematical models of Horton and Strahler related to the drainage network may not be applied in totality but the linear relationship stands valid.

The paper by Padmaja and Soundarvalli discusses the behavioural pattern of bifurcation ratios in the Godavari and Krishna basins. The authors have noted the variations of bifurcation ratio in these two basins and attributed this variation in Rb to varying geological formations and climatic conditions.

SUGGESTED READINGS

Clarke, J.I., 1966 : Morphometry from maps, *Essays in Geomorphology*, Edited by Dury G.H., New York, 235-274.

Chorley, R.J., 1966 : The application of statistical methods to geomorphology, *Essays in geomorphology*, Edited by Dury G.H., New York, 275-387.

Chorley, R.J., 1969 : *Water, Earth and Man*, Methuen : London.

Gardiner, V., & Park G.C., 1978 : Drainage basin morphometry : review and assessment, *Progress in Pysical Geography* 2, 1-35.

Gardiner, V., 1973 : The univariate distributional characteristics of some morphometric variables, *Geografisker Annaler Series A 54A*, 147-53.

Horton, R.E., 1932 : Drainage Basin Characteristics, *Trans. Am. Geophysics*, 13, 350-61.

Horton, R.E., 1945 : Erosional Development of streams and their drainage basins : Hydrophysical approach to quantitative morphology, *Bull. Geol. Society Am.* 56, 275-370.

Kesseli, J.F., 1946 : A neglected field of geomorphology, *Annals of Association of American Geographers*, 86, 93.

King, C.A.M., 1976 : *Landforms and geomorphology : Concepts and History*, Benchmark papers in Geology, 28.

Leopold, Wolman, Miller, 1964 : *Fluvial Processes in Geomorphology*, Eurasia Publishing House Pvt. Ltd., New Delhi.

Miller, V.C., 1953 : *Relation of Quantitative Geomorphic study of Drainage Basin characteristics in the Clinch Mountain Area*, Virginia and Tennesse, Technical Report 3, Office of the Naval Research, Dept. of Geology, Columbia University, New York.

Melton, M.A., 1958 : Correlation structure of morphometric properties of drainage systems and their controlling agents, *Journal of Geology*, 66, 442-460.

Morisawa, M.E., 1962 : Relation of quantitative geomorphology of stream flow in representative watersheds of Appalachian plateau province *Bull. Geol. Soc. Am.* 73, 1025-1046.

Maxwell, J.C., 1960 : Quantitative geomorphology of the San Dimas experimental forest, California, Office of Naval Research, Geography Branch, Project NR 389-042, *Technical Report* 19.

Robinson, G., 1963 : A consideration of the relation of geomorphology and geography, *Professional Geographer*, 12, 5-15.

Schumm, S.A., 1956 : Evolution of Drainage systems and slopes in Badlands at Perth Ambsy, New Jersey, *Bull, Geol. Soci. Am.* 67, 597-646.

Scheidegger, 'A.E., 1967a : On the topology of river nets, *Water, Resource Research*, 3, 103-106.

Strahler, A.N., 1952 : Quantitative geomorphology of Erosional landscapes, *19th International Geological Congress Algiers*, Section 3, Vol. 3, 341-354.

Strahler, A.N., 1958 : Dimensional analysis applied to fluvially eroded landforms, *Bull. Geol. Soc. Am.*, 64, 279-300.

Smith, K.G., 1958 : Erosional processes and landforms in Badlands, National Monument, S. Dakota, *Bull. Geol. Soc. Am.* 69, 975-1008.

Shreve, R.L., 1966 : Statistical law of stream numbers, *Journal of Geology*, 74, 17-37.

Walsh, F., 1972 : Application of stream order numbers to the Merrimark river basin, *Water Resource Research*, 8, 141-144.

Ward, R.C., 1975 : *Principles of Hydrology* (2nd Edi.) McGraw Hill, London.

ONE

Goutam Bhattacharya and *V. K. Verma*

SIXTIES' PARADIGM IN GEOMORPHOLOGY

IN THE LIGHT of Thoman Kuhn's (1962) concept the year 1945 is essentially a major historical landmark in the growth and development of geomorphological thoughts. Robert E. Horton provided a new paradigm in geomorphology : pioneered the quantitative approach to drainage basin description realizing that the hydrologic characteristics of a drainage system are organically linked to their morphologic characteristics. He analyzed the functionality of their inter-relationship in quantitative terms. His 1945 paper provided the structure and direction for the geomorphological research that followed. The seed of this structural-functional analysis was sown at a time when even the concepts of systems approach were not in vogue in geology on geography. This simplistic structural-functional analysis by Horton attracted a host of his followers who were better equipped with the tools of mathematics. This brought about a revolution in geomorphology.

Horton (1945) for the first time introduced the concept of stream orders. First order streams were considered as those that have no tributaries, and a second order stream as those that receive only streams of the first order as tributaries. The main stream is denoted by the same ordinal number all the way to its headwaters by renumbering one of the longest or the most direct first order stream as second order. The procedure is repeated with higher order streams, so that, n^{th} order stream extends headwards to the beginning of the longest tributary, through the process of renumbering one of the lower order streams every time two channels of equal order joined. Strahler (1957) considered that at the source of a river, the order is by definition equal to 1. The order changes by a factor +1 whenever the stream is joined by another of the same order. The order of a particular stream in this method is always denoted by an integer number. A stream with lower order m joins another stream with a higher order n produces no change in the order of the resultant stream. Scheidegger (1966b) expressed in mathematical terms the composition law of stream order numbers which is as follows :

$$n^*m = max\ (n, m) \text{ for } n \neq m$$
$$= n+1 \quad \text{ for } n = m$$

The possible orders of any stream segment in a basin are all the integers $n \leq N$, when N is called the order of the basin.

The criticism that was labelled against Horton-Strahler system of ordering was that the ordering scheme is very insensitive to variations in structure and lithology. Bifurcation ratios were found to be remarkably stable from one area to another and generally cluster in the range 3.5 to 4.0 (Werrity, 1972). The Horton-Strahler stream ordering system does not recognise the junction of a stream of order n with a lower order stream; causing unsatisfactory description of the topology of a river network (Scheidegger, 1966b).

The inconsistencies in Horton-Strahler ordering system were responsible for the beginning of another trend—the application of topology in geomorphology. In an attempt to generate a more sensitive ordering scheme devoid of Horton's inconsistencies, Scheidegger (1965) proposed a consistent law of composition of stream numbers n and m by the following formula :

$$n^*m = \log\ (2^n + 2^m) \ / \ \log 2$$

In Scheidegger's consistent scheme of stream ordering, ordering can be denoted by positive integers, enabling distinct generation of river segments applicable in Strahler's scheme as well. However, Fraser Walsh (1972) points out that orders greater than two take non-integer values although it is possible to express these segment orders as even integers I when :

$$n = \log\ (I_1) \ / \ \log 2$$

The integer order system sets all sources as having an integer value of 2 (n=1) and order integers are added whenever a junction occurs, meaning thereby, that the basin integer order is equal to twice the number of stream sources. Therefore,

$$b = \log (\Sigma I_n)/\log 2,\ n=1$$

The problem centres on the accuracy in estimating ΣI_1. Fraser Walsh having pointed the preceding problems went on to make an amendment to the Scheidegger ordering system based on the length of stream source segments. This he did by changing the order of stream sources by the ratio of the length of the source segment Si to the observed mean length $\overline{L}$ of source segments; S_i and $\overline{L}$ were measured and calculated respectively. The order of sources was then calculated by the following expression :

$$n_i = \log (2.\ S_i/\overline{L})/\log 2.$$

These amended orders enabled identification of incorrectly mapped sources and helped to identify which junctions in a single Scheidegger order have longer than average sources upstream.

A vigorous search for a suitable branching process theory that could act as an analog model making possible the theoretical deduction of Horton's laws of stream and ultimately construct an acceptable 'river theory' remained a major concern of practising geomorphologists in the midsixties. Scheidegger 1965, 1966 a, b, 1967 a, b; Ranalli and Scheidegger 1968 a, b; Smart 1967, 1969; Sharp 1971; Walsh 1972; explained network development in terms of a stochastic branching process and proposed random topology model based solely upon combinatorial properties. Shreve (1966) proposed a model based on random connections in a topological network.

From these initial formulations developed a new system of stream ordering based largely on Scheidegger (1965) and Shreve (1967). Each section of a stream in a channel was now numbered in terms of the number of finger tip tributaries or sources. Each finger tip tributary is assigned a magnitude 1. Each of these tributaries is a section of a channel called an exterior link that starts at a point called a junction or fork. Whenever two exterior links join to form a fork the downstream section of channel called an interior link is assigned a magnitude 2. Whenever any two links whether interior or exterior having magnitudes M_1 and M_2 join the resultant downstream interior link has magnitude M_1+M_2. Now at this stage Leopald and Langbein's (1962) random walk model assumed a greater significance : it paved way for the generation of statistical data and analysis of drainage network relations. It was found that the concept of magnitude of channel links and numerical relationship with their upstream and downstream neighbours satisfactorily fitted into the topological random model which allowed the exact calculation of the probability of occurrence of each link type for random population of networks of any magnitude (Mock, 1971).

Liao and Scheidegger (1969) presented tables of calculations of the expected stream lengths and drainage areas of Strahler streams of various orders for networks with various number of first order streams, based on the random-graph model of such networks. Seginer (1969) presented a random walk and random roughness models of drainage networks and Howard et al., (1969) characterized braided streams. Coffman and Turner (1971) made use of computer to determine the geometry and topology of stream networks based on binary string procedures. By this time a number of binary string method for describing the topology of stream networks came into existence. Binary string method had restricted uses, although they were considered an efficient tool for describing network topology. There was another advantage : binary notation provided efficient computer information storage. The binary string method of Scheidegger (1967), Shreve (1967), Liao and Scheidegger (1968) and Smart (1970) did not assist geomorphic and engineering studies that require additional non-binary strings. Coffman and Turner (1971) made four pertinent observations that led them to propose a new system creating scope for using additional non-binary informations of networks. The four observations are as follow :

1. Liao and Scheidegger's (1968)* algorithm for computing Strahler stream orders for binary strings fails to analyse all the laws of stream numbers proposed by Horton (1945) since they involve non-topologic properties such as length and gradient.

* Cited in Coffman and Turner (1971).

2. Liao and Scheidegger (1968) defined the topology of stream networks in terms of junction and source points. In case, if more than two types of points are used the binary notation must be sacrificed. Furthermore, if this technique is used, strings defining link properties do not have one to one correlation with the original binary strings.
3. Smart (1970) suggested additional non-binary strings containing information on network properties such as length and gradient. Such strings were developed separately and externally offsetting the advantages of computer.
4. Smart's (1970) method also had the disadvantage that it did not provide an easy solution to define specific control points within a network. He, instead, gave a cumbersome and error prone method wherein he suggested that these points be defined by adding fictitious junctions and first order of exterior links to the network.

As a reaction to these restrictions Coffman and Turner (1971) proposed the water system concept by using both co-ordinate data describing the point geometry and topologic data describing network links. They extended the Liao and Scheidegger method to define the topology and later converted the topologic information to refer to the links. This they achieved by collecting points in a clearly defined sequence from mouth to source and defined the purpose of each point by a suitable code. This method facilitates computation involving area, length or perimeter.

The bifurcation ratio which was abundantly used came to be suspected as a viable method of investigating topologic structure of networks. This ratio although gives useful information about individual networks, is generally ineffective in distinguishing differences in network structure due to lithologic control and degree of maturity (Smart, 1972). This failure is attributed to two factors :

1. Random nature of network topology.

2. This ratio is calculated from average values that dilute many details which could otherwise throw light on lithologic control and degree of maturity.

Coffman et al., (1972) developed a new topologic relationship as an indication of drainage network evolution. The relationship between the number of links and the number of segments of a network was envisaged as a curve with the general form :

$$Y=2X-(2^n-1)$$

Where Y is the number of links, X is the number of segments and n is the Strahler stream order of magnitude 2, 3, 4 or 5. They used the curve to delineate threshold and hypothetical boundary conditions in order to predict stream order. In this paper they also made another important suggestion that for any network of given order, the bifurcation ratio increases to an improbable value. When this value is reached, branching increases the order of the network, and then the bifurcation ratio is decreased.

In the same period Jarvis and Smart made two important contributions : both of them included duality between points and lines along with the statistical geometric properties of network. Jarvis (1972) redefined the concept of link distance of Scheidegger (1965) and choosing and redefining two parameters—H_{max} and H_e of Liao and Scheidegger (1968) developed a time dependent measure of topologic structure of the network called E index, where :

$$E=\Sigma M.\ H_i \ / \ \Sigma M.\ H_e$$

M is the magnitude of a given point and H is its link distance. The subscript i and e denote summation over the interior points and over the exterior point sources respectively. Jarvis in this paper claimed "The E index is a more precise structural measure of incorporating all the topologic information contained in the network graph". The M and H parameters summarise drainage development and structural configuration of drainage network respectively.

Smart (1972) constructed four dimensionless indices such as :

$$\lambda = l_e/l_i$$

$$\propto = a_e/a_i$$

$$K_e = l_e 2/a_e$$
$$K_i = l_i 2/a_i$$

where l_e—exterior link length

l_i—interior link length

a_e—mean area associated with exterior links

a_i—mean area associated with interior links.

These data drawn from various geologic population indicated that these indices are effective in detecting differences due to varying lithology and degree of maturity.

In summing up it may be said that the Hortonian revolution undoubtedly gave rise to a new paradigm, the structure of which may be resolved into the following parts in a chronological heirarchy :

1. A pre-paradigm stage, the history of which has not been satisfactorily linked up to the year 1945.
2. Quantitative approach to drainage basin approach; linkage of morphologic and hydrologic characteristics; introduction of structural-functional analysis; and formulation of laws and theories of river behaviour obeying the grammar of mathematics.
3. Identification of inconsistencies and efforts to prove it; search for a new approach to amend it.
4. Leopald and Langbein's (1962) random walk model of stream network which they exploited heuristically by generating random sequences on a computer.
5. Cognisance of the role of binary mathematics.
6. Applications of topology, graph theory alongwith application of probability statistics.
7. The algebraic treatment of topological data helped remove fundamental drawback of ordering in Horton's system; statistics brought many revealing facts to the surface. Horton's laws were modified with greater mathematical sophistication.
8. Marriage of geomorphology with hydrology required many other informations which a binary system was incapable of proper handling.
9. New indices were constructed with the help of topological geometric variables of river networks. This had a dual effect : first of all it widened the applicability of

geomorphology in water resource studies; and second the statistical properties of geometry and structure of networks provided an acceptable alternative to the bifurcation ratio and helped to evaluate drainage evolution in a better way.

CONCLUSION

The main purpose to review the geomorphological literature of the sixties and the early seventies was to reassert that Horton's 1945 contribution has been revolutionary; and it is so for two reasons :

1. It sowed the seed for structural-functional analysis because of quantification of many geomorphic variables. This quantitative approach metamorphosed the discipline of geomorphology. This development had another great effect : it brought geomorphology closer to many other disciplines making it, in a real sense, multidisciplinary in nature.
2. The inconsistencies in Horton's method itself were of tremendous epistemological significance. The Hortonian inconsistency facilitated entry of topology in geomorphology. Application of topology not only elevated the status of this discipline, it also brought geomorphology closer to hydrology and water resource studies. The long awaited marriage of geomorphology with hydrology widened social relevance of geomorphology was greatly lacking in geomorphology of pre-second world war years.

The year 1945 is then a landmark in the history of geomorphology. This year initiated genesis of a new paradigm which blossomed in the sixties. An enquiry into the pre-paradigm stage has not been possible due to lack of adequate historical material. A passing reference to two different works of Gravelius (1914) and another that of Horton (1932) may be made here that may give some clue to reconstructing the history of the pre-paradigm stage. These two clues may not be adequate. In fact, it is our conviction that a mine of information in this context may be available from the military archives of those countries who were involved in the two world wars. In case if no such records are available the only solution left would be the retrieval of experiences from those geologists and geographers who were engaged in military intelligence in the two world wars and the intervening period.

SUGGESTED READINGS

Coffman, D.M. and Turner, A.K., 1971 : Computer determination of the geometry and topology of stream networks, *Wat. Res. Research*, 2, 419-423.

Coffman, D.M.; Keller, E.A., and Melhorn, W.N., 1972 : New topologic relationship as an indicator of drainage network evolution, *Wat. Res. Research*, 8:6, 1497-1505.

Gravelius, H., 1914 : Flusskunde Berlin, G.J. Goshen.

Horton, R.E., 1932 : Drainage basin characteristics, *Trans. Am. Geophys.*, 13, 350-61.

Horton, R.E., 1945 : Erosional development of stream and their drainage basins : Hydrophysical approach to quantitative morphology, *Bull. Geol. Soc. Am.*, 56, 275-370.

Howard, A.D.; Keetch, M.E. & Vincent, C.L., 1969 : Topological and geometrical properties of braied streams. *Wat. Res. Research*, 6:6, 1674-1688.

Jarvis, R.S., 1972 : New measure of topologic structure of dendritic drainage networks, *Wat. Res. Research* 8:5, 1265-1271.

Kuhn, T.S., 1962 : *Structure of Scientific Revolution*, Chicago University Press.

Leopald, L.B. and Langbein, W.B., 1962 : The concept of entropy in landscape evolution, U.S., *Geol. Surv. Prof. Paper 500-A*.

Liao, K.H. and Scheidegger, A.E., 1969 : Theoretical stream length and drainage areas in Horton nets of various orders, *Wat. Res. Research*, 563, 744-46.

Mock, S.J., 1971 : A classification of channel links in stream networks, *Wat. Res. Research*, 7:6, 1558-66.

Ranalli, G., and Scheidegger, A.E., 1968a : Topological significance of stream labelling methods, *Bull. Int. Assoc. Sci. Hydrol.*, XIII : 3-4, 77-85.

Ranalli, G. and Scheidegger, A.E., 1968b : A test of the topological structure of river nets, *Bull. Int. Assoc. Sci. Hydrol.*, xiii : 2, 142-153.

Scheidegger, A.E., 1965 : The algebra of stream order number, U.S. Geol. Surv. Prof, *Paper 525-B*, 187-189.

Scheidegger, A.E., 1966a : Stochastic branching process and the law of stream numbers, *Wat. Res. Research*, 2:2, 199-203.

Scheidegger, A.E., 1966b : Statistical description of river networks, *Wat. Res. Research*, 2:4, 785-790.

·Scheidegger, A.E., 1967a : On the topology of river nets, *Wat. Res. Research*, 3:1, 103-106.

Scheidegger, A.E., 1967b : Random graph patterns of drainage basins, *Paper 157*, 14th Gen. Assembly, U.S.G.I., Int. Assoc. Sci. Hydrol., Beme.

Seginer, I., 1969 : Random walk and random roughness models of drainage networks, *Wat. Res. Research*, 5:3, 591-607.

Sharp, W.E., 1971 : An analysis of the law of stream order for Fibonacei drainage pattern, *Wat. Res. Research*, 7:6, 1548-1557.

Shreve, R.L., 1966 : Statistical law of stream numbers, *Journ. of Geol.*, 74, 17-37.

Shreve, R.L., 1967 : Infinite topologically random channel networks, *Journ. of Geol.*, 75. 178-86.

Smart, J.S., 1967 : A comment on Horton's law of stream numbers, *Wat. Res. Research*, 3, 773-6.

Smart, J.S., 1969 : Topological properties of channel networks, *Bull. Geol. Soc. Am.*, 80, 1757-74.

Smart, J.S., 1970 : Comment on dependence of stream link length and drainage on order, *Wat. Res. Research*, 6, 1424.

Smart, J.S., 1972 : Quantitative characterisation of channel network structure, *Wat. Res. Research*, 8:6, 1487-1497.

Strahler, A.N., 1957 : Quantitative analysis of watershed geometry, *Trans. Amer. Geophys. Union*, 38, 913-20.

Walsh, F., 1972 : Application of stream order numbers to the Merrimack river basin, *Wat. Res. Research*, 8:1, 141-144.

Werrity, A., 1972 : The topology of stream networks, in R.J. Chorley (ed.), *Spatial analysis in Geomorphology*, Methuen & Co., London, 167-197.

TWO

V. Gardiner

THE IMPACT OF CLIMATE ON FLUVIAL SYSTEMS

THE NATURE of river network must inevitably depend to some extent upon climate since the geometry of river channels is controlled by discharge, which depends upon the balance amongst all components of the hydrological cycle, not least of which are the climatic elements. Rainfall input to the hydrological system is modified by temperature, humidity and wind conditions, and the totality of such effects, which we envisage as climate, is manifested in terms of output from the hydrological system as runoff and evapotranspiration losses. An understanding of the relationships between fluvial geomorphology and climate, therefore, necessitates some acquaintance with the elements of the hydrological cycle, as summarised in texts such as Ward (1975) and Chorley (1969).

One of the most significant aspects of climate with respect to fluvial systems is the seasonal distribution of meteorological elements or climatic regime, and in particular the effective rainfall regime. Networks which experience identical annual rainfall inputs may differ widely in the way in which the rainfall input and/or evapotranspiration losses are distributed temporally. Various indices of seasonality have been proposed and attempts made to devise universal systems of regime classification. The most commonly used is that of Pardé (1933), who recognised three main types of river regime (Figure 2.1). Simple regimes are those which have only one period of high runoff and one period of low runoff. These may be produced by a variety of causes. For example, in the higher latitudes of the northern hemisphere a high water period in July and August, caused by melting of glacier ice, may be followed by low winter runoff when temperatures are low and ice melt is negligible. Alternatively, in similar latitudes, an ice-free area of fairly evenly distributed rainfall may have a fall in runoff in summer months when evapotranspiration losses are at a maximum. In the tropical regions evapotranspiration losses are more nearly evenly distributed throughout the year, and rainfall is the main determinant of the hydrological regime.

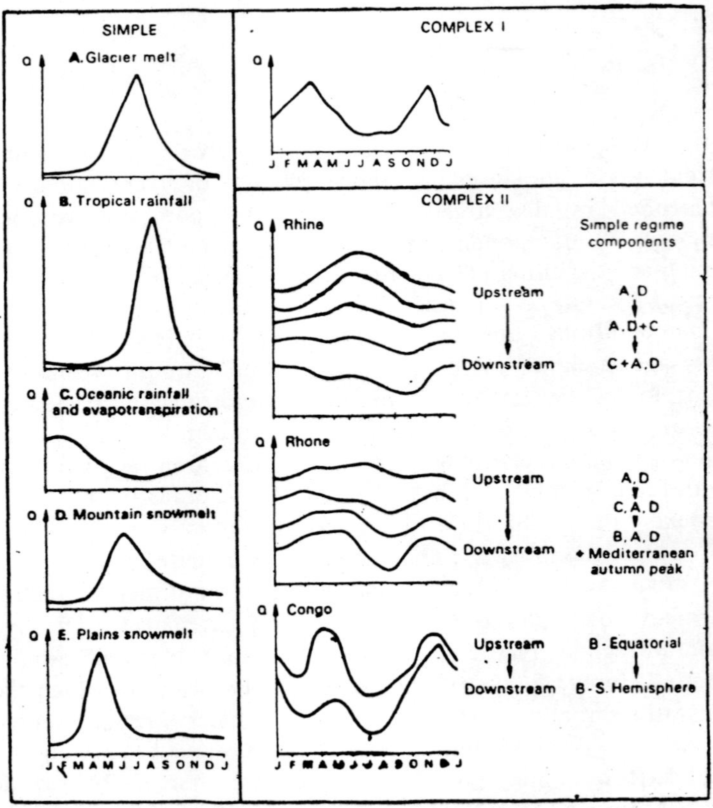

Fig. 2.1 : *Generalised examples of simple and complex regimes* Following Pardes (1933) classification. Discharge (Q) is the ratio of each monthly mean value to the yearly mean.

Complex I regimes have at least four and sometimes as many as six identifiable hydrological phases. These may again result from a variety of causes but one example might be spring snowmelt runoff being followed by a second runoff peak in summer due to conventional rainfall, or in winter due to reduced evapotranspiration losses.

Complex II regimes are probably the most common type, and are characteristic of most of the larger rivers of the world. Complex I regimes may occur in the headwaters of major river systems but these become progressively modified downstream as rivers flow through varied terrains and climatic regions, and receive tributaries with differing regimes. One variant is the glacier or snowmelt fed stream in the upper reaches which becomes increasingly modified downstream by a single type of rainfall regime. A second type of Complex II regime is similar to that described above except that it is modified by at least two types of rainfall regime. The third type of Complex II regime is the type influenced only by rainfall regimes, but which combines characteristics from two or more rainfall zones.

Such categorisations of runoff regime are of great significance to studies of fluvial networks, as river channels must necessarily accommodate the range of flows likely to be experienced, and the range of flows depends upon the river regime. The regime concept can also be applied to sediment concentrations in rivers, as illustrated by, for example, Gregory and Walling (1973, Fig. 6.13), but sufficiently detailed and accurate information to permit this is rarely available.

Whilst climate is undoubtedly of significance in determining the nature of fluvial systems, it should be borne in mind that rivers do not operate in isolation, but form part of the entire spectrum of geomorphological processes. Many attempts have been made to erect classifications of geomorphological processes and landforms with respect to major climatic regions, and such attempts at morphoclimatic classification are reviewed by Stoddart (1969) and Derbyshire (1976). These have been further developed into morphogenetic classifications (e.g. Budel, 1963) in which temporal changes in morphoclimatic regime are incorporated. Fluvial processes should inevitably form a strong component of such attempts, however, attention has perhaps too often been diverted to the more esoteric landforms produced by aeolian activity and periglacial processes, whilst there has been, to varying degrees, an unfortunate assumption that fluvial activity is 'normal' and unvarying. Some preliminary indications of varying intensity of fluvial processes according to morphoclimatic regime have, however, been given (e.g. Peltier, 1950) and in a valuable collection of essays (Derbyshire, 1976) a number of workers have illustrated this theme. Gregory (1976) presented a comprehensive review of data on network geometry and climate; Morgan (1973, 1976) analysed data on drainage texture in relation to climatic variation, and the variation of drainage density with climate has been examined by Melton (1957) and Gregory and Gardiner (1975). Despite these attempts the climatic dimension to fluvial variability is one which has not yet been fully explored. The development of climatically based process-response models is hindered by the lack of information on specific regions. Whereas much morphometric discharge and sediment data are now available for many temperate areas, there is a paucity of information from tropical and polar regions, where such data as are available are often either of limited duration or are of a purely reconnaissance nature. This brief review of some climatic influences on drainage networks must, therefore, necessarily reflect in its emphases the differing amounts of information available. It examines climatic influences on river systems at several levels of spatial organisation or hierarchical complexity, starting with the simplest fundamental unit where runoff is generated, the slope. Channel cross sections and reaches are considered and finally entire drainage basins and networks are examined.

In recent years the mechanisms by which runoff is generated on slopes have been the focus of much research, since an understanding of these processes is fundamental to all hydrological response and, hence, to the nature of fluvial networks. Hillslope hydrological processes have been broadly categorised by Dunne (1976) as in Fig. 2.2, from which it can be seen that the occurrence of the three major groups of runoff type is conditioned by two groups of controls, one of which encompasses a direct climatic element. Although any or all of the three groups of runoff types may occur in a particular area, the dominant process is at least partly directly controlled by climate. In arid and semi-arid climates Hortonian overland flow dominates, since rainfall intensities when precipitation does occur are usually greater than infiltration capacities of the soil surface, which is often baked and compacted. In more humid climates the opposite relationship whereby rainfall

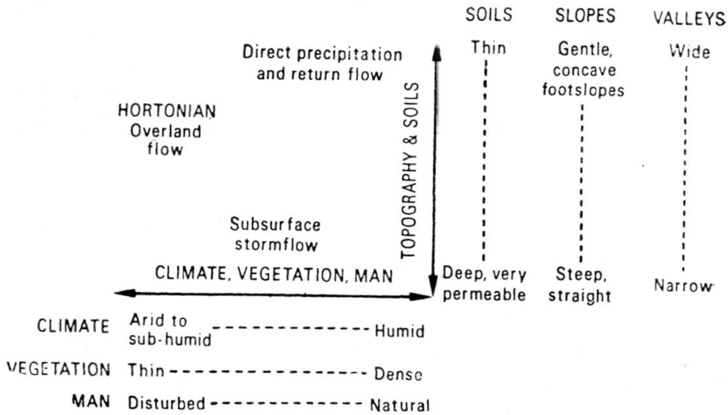

Fig. 2.2 : *Schematic model of controls of three generalised runoff types onslopes slopes* (*after Dunne, 1976*).

intensities are less than infiltration capacities is more common, and runoff is, therefore, dominated by subsurface storm flow, direct precipitation flow or return flow, depending on the topography and soils of the individual slope. In addition, of course, climate's impact is also manifested via its influence upon both vegetation and soil type, and of paramount significance is their effect upon the development of soil organic matter and, hence, infiltration capacity of the soil.

Despite the existence of this generalised division of dominant runoff processes between humid and non-humid climates, it is difficult to identify the values of climatic variables which act as thresholds between the flow type dominance regions, since not only are many other environmental characteristics of significance, but climate itself is a compound concept embracing precipitation amount and intensity, temperature and evaporation. However, Kirkby (1976) has examined some of the limited relevant data. The relationship which he identifies between a topographic factor and overland flow occurrence is depicted for four regions of diverse climatic types in Figure 2.3 and Table 2.1. In semi-arid S.E. Arizona overland flow is produced by the Hortonian mechanism, and is, therefore, independent of the topographic factor, occurring ubiquitously. In the other regions overland flow generation is related to the topographic factor since it is produced at least partly by non-Hortonian mechanisms. In the United Kingdom high rainfall case overland flow can readily be generated on straight slopes, whereas in the other two regions it can only occur in definite hollows. The slopes of the lines in Fig. 2.3 are determined mainly by the mean rate of water entry into the soils, and a high sensitivity to topography is favoured by high rainfall and relatively low evapotranspiration. Thus, in cool wet climates overland flow is very sensitive to topography

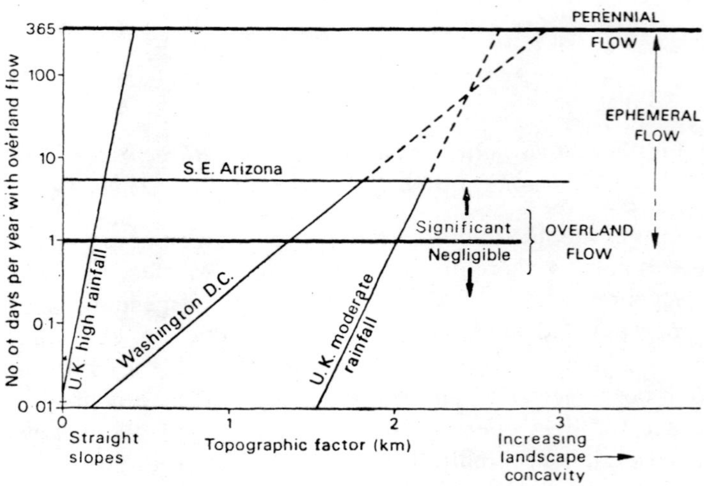

Fig. 2.3 : *Relationship between overland flow occurrence and a topographic factor for four sets of data*

The topographic factor as a/s, where a=drainage area per unit length of contour and s=slope. (After Kirkby, 1976 with additions.)

whereas most humid temperate areas demonstrate a lesser degree of sensitivity. Kirkby suggests that as a first approximation appreciable overland flow and, hence, appreciable erosion and sediment transport may be considered to occur when overland flow frequency exceeds one day per year. Thus, critical thresholds for the occurrence of this minimum value of wash processes may be identified, as in Table 2.1; similarly he also suggests that similar considerations may govern the maximum occurrence of overland flow, at which frequency flow becomes perennial, thus, leading to a consideration of water in river channels. Values of the topographic factor for perennial overland flow are also given in Table 2.1, where again a climatic influence is evident. The zone having values of the topographic factor between the two limits given may be envisaged as a zone of ephemeral overland flow, which gives rise to the dynamic nature of drainage density as illustrated elsewhere (Figure 2.1). Such considerations afford a possible aid to the definition of the extent of the functioning drainage net, which is of obvious significance in the determination and use of drainage density (e.g. Gregory and Gardiner, 1979),

TABLE 2.1 : **Data for areas examined in Figure 2.3**

	U.K. moderate rainfall	U.K. high rainfall	Washington D.C., U.S.A.	S.E. Arizona U.S.A.
Annual rainfall (mm)	700	2500	1000	300
Mean rainfall per rainday (mm)	4	10	10	15
Average infiltration intensity over year (mm/day)	1	5	1	0
No. of raindays per year	175	275	100	20
Critical topographic factor for overland flow ⩾1 day/year	2 0	0.2	1.3	0
Critical topographic factor for perennial overland flow	2.5	0.5	5.0	00
Predicted minimum drainage density (km/km^2)	0.3	5.6	1.4	180

The topographic factor = a/s

where a = drainage area per unit length of contour

s = slope

and may also be used (Kirkby, 1976) in the estimation of predicted minimum drainage densities (Table 2.1). The above considerations are, however, supported by relatively little field data, and concern only overland flow processes. However, sufficient observations are available to suggest that in general terms the processes of runoff generation on slopes do differ in different climatic regions. The geographical distributions of observations is, however, uneven and whereas much information is now available from temperate regions, the same is not true elsewhere.

One zone in which runoff generation is possibly least well understood is the Arctic region, which is characterised by extremely short runoff seasons, of perhaps only two or three months in the year e.g. McCann and Cogely, (1973). Wilkinson and Bunting (1975) have examined the processes of rill discharge and sediment transport in an Arctic region, and concluded that considerable variation may exist in the amount of sediment transported in particular rills. Some may carry little or no load, being fed from moss-covered or vegetated areas, whilst others may carry considerable loads. Snow patches which feed rills may act as sediment traps, since they tend to be located at the edge of the interflow zone, at a break of slope, and temporal variations of sediment transport may occur as a response to the exposure of sediment in the melting snow. The importance of snow patches to slope runoff has also been documented by Bllantyne (1978), who made careful measurements of rill discharge and of snow patch water equivalent, and concluded that discharge was not, as has often been argued, fed entirely by snow melt. Instead he suggested that the snow patch occurs at a break of slope where saturated throughflow becomes overland flow.

There has also been considerable debate concerning the nature of flow generation processes in tropical areas, and only recently have sufficient instrumental observations become available to allow some judgements to be made. One recent study is that of Gilmour and Bonell (1979), who reported results from catchments under tropical rainforest in Australia. They pointed out that rainfall has a far greater intensity than in temperate areas, and there are ample visible indications of overland flow, as reported similarly elsewhere in the tropical rainforest (e.g. Ruxton, 1967; Löffler, 1977). Overland flow volumes are high and lag times are characteristically short. Throughflow tends to be restricted to the upper soil horizons with little deeperflow. At the cessation of flow the overland flow contracts, and a short period of throughflow follows. Because the soils are perpetually near saturation the antecedent conditions are unimportant to runoff generation (Morgan, 1972), and drainage densities are high, with the variable source area concept consequently being less important than in temperate areas. This may be contrasted with, for example, the study of Gurnell (1978) which demonstrates excellently the dynamic nature of a temperate drainage network, although she was unable to isolate any significant influence of antecedent conditions upon drainage density in this particular case.

It should, however, be noted that these general conclusions may not necessarily apply to all locations. For example, Morgan (1972) reported that in a gully catchment in Malaysia overland flow was relatively limited in significance, with erosive flows in the gullies having a recurrence interval of about 60 days. Throughflow appeared to be more important than overland flow, and was reported on four occasions during eight months of observation, at the basal weathering front 1 m beneath the ground surface.

Despite the limited amount of observations available examples such as these clearly suggest that climate influences methods and rate of runoff generation, and since drainage channels and networks are adjusted to runoff, climate must also influence the nature of fluvial networks. The exact nature of this relationship and the processes from which it results are, however, less clear. On the microscale it has been suggested above how climate influences network dynamics in terms of network expansion and contraction as adjustment to precipitation. On a larger scale Abrahams (1980) has recently suggested how a link may exist between microscale hillslope hydrological proeesses and macroscale drainage density. A continuum is envisaged from Hortonian overland flow to saturated throughflow as the dominant stream flow generation process; in landscapes with the former type of flow generation the depth of overland flow is proportional to the source-area size, and drainage density should be positively correlated with ground slope. In landscapes in which saturated overland flow and subsurface flow are dominant source-area size is not related to ground slope, and drainage densities are, therefore, independent of ground slope. The nature of the relief-drainage density relationship is, therefore, expected to differ according to climatic region. At one extreme a climatic regime which gives rise to a densely vegetated thick soil and regolith will generate runoff by saturated throughflow and saturated overland flow, and drainage density will be largely independent of slope aspects of the basin. At the other extreme a climate giving rise to a sparsely vegetated basin with thin soils will generate Hortonian overland flow predominantly, and drainage density will depend upon slope of the basin. The increasing strength of relationships between drainage density and slope aspects of basins with increasing dominance of overland flow is supported by data in Table 2.2.

The considerations and examples presented above serve as sufficient indication that at the most basic level of network operation climate exerts an influence upon network characteristics, and lead on to a consideration of river channels. The geometry of a river channel cross-section is adjusted to the range of flows experienced by that section, with, according to Dury (1969) the bank-full flow having a recurrence interval of 1.58 years. The bank-full flow is suggested to be a particularly significant one in terms of the channel geometry, although more recent work (e.g. Kennedy, 1972) suggests that bank-full discharge may have a wide range of values of recurrence interval, and that channel-forming discharge is necessarily a range of flows rather than a particular value. Since the range of flows and the recurrence interval of flows is controlled by climate it is, therefore, to be expected that some degree of variation of channel geometry according to climate may occur, although the fundamental physical laws governing the flow of water, transport of sediment and bank resistance remain essentially unaltered, whatever the climate.

One way in which the geometry of river channel cross-sections has been described is by means of the hydraulic geometry approach pioneered by Leopold and Maddock (1953). In this the morphology of channel cross-sections is described by two basic hydraulic exponents, which express the rates at which

TABLE 2.2 : **Some relationships reported between slope and drainage density aspects of drainage basins**

Study	Variables and analysis	Area	Observed flow type	Results
Abrahams 1980	Link density; relief ratio. Rank correlation	Perth Amboy U.S.A.	Hortonian overland flow	Very significant correlation, $p<0.001$
Gregory 1976	Valley density; relief ratio and slope	Alaska	Overland flow?—ephemeral net	Significant correlation
Gardiner 1972	Drainage density; relief ratio. Product moment correlation, transformed data	Culm Measures Devon England	Subsurface flow dominant but saturation flow wide-spread	Significant correlation $p<0.01$
Gregory 1976	Stream density; relief ratio and slope	Alaska	Throughflow?—perennial net	No significant correlation
Gardiner (unpublished)	Drainage density; relief ratio. Rank correlation	Ardennes, Belgium, as in Fig. 2.2 of chap. in this volume	Subsurface flow dominant, saturation flow spatially restricted	No significant correlation $p>0.25$
Abrahams 1980	Link density; relief ratio. Rank correlation	Dartmoor, England	Subsurface flow dominant, saturation flow spatially very restricted	No significant correlation $p=0.67$

width and depth change with increasing discharge. Similar exponents may also be derived for at-a-section changes in process rates such as velocity and sediment transport, as well as for other morphological measures such as capacity (cross-sectional area) and slope. Park (1977) has presented an analysis with respect to climate of some of the at-a-station exponents reported in the literature. He concluded that although there appeared to be some slight differences between the pattern of exponents reported from four different climatic regions, the impact of climate was small. Rhodes (1977) has similarly suggested that factors such as channel pattern and sediment type are of greater significance than climate. However, Park does suggest that humid temperature streams have a tendency (Fig. 2.4) towards low to medium width and medium depth and velocity exponents, whereas his limited data suggest that tropical channels tend to have low width and high depth exponents. There is, however, considerable scatter about these general trends, and there are many other factors which need to be considered.

For example, Gupta (1975) examined the at-a-station hydraulic geometry of two streams in Jamaica and suggested that seasonality of rainfall is responsible for the differences between them. The stream with a markedly seasonal regime is characterised by a wide, shallow, flat-bottomed braided channel with steep banks, whereas the stream with a more uniform regime has a meandering river with a deep round-bottomed channel with sloping banks.

One unusual control of at-a-section channel geometry by climate is that identified by Smith (1979) for Arctic rivers. He established that the recurrence interval of bank-full flows for rivers in Alberta, Canada, was of the order of 10-20 years, which may be compared with recurrence intervals for bank-full flows of about 1-2 years for rivers in temperate climates. This apparently gross inflation in channel dimensions, by comparison with the flows experienced, he suggested was due to two processes not experienced under other regimes. First, ice jams caused by drifting ice during the spring break-up period caused considerable ponding back of water upstream, and enlargement of the channel to accommodate this. Secondly, anchor or frazil ice attached to the river bottom acted as an additional process

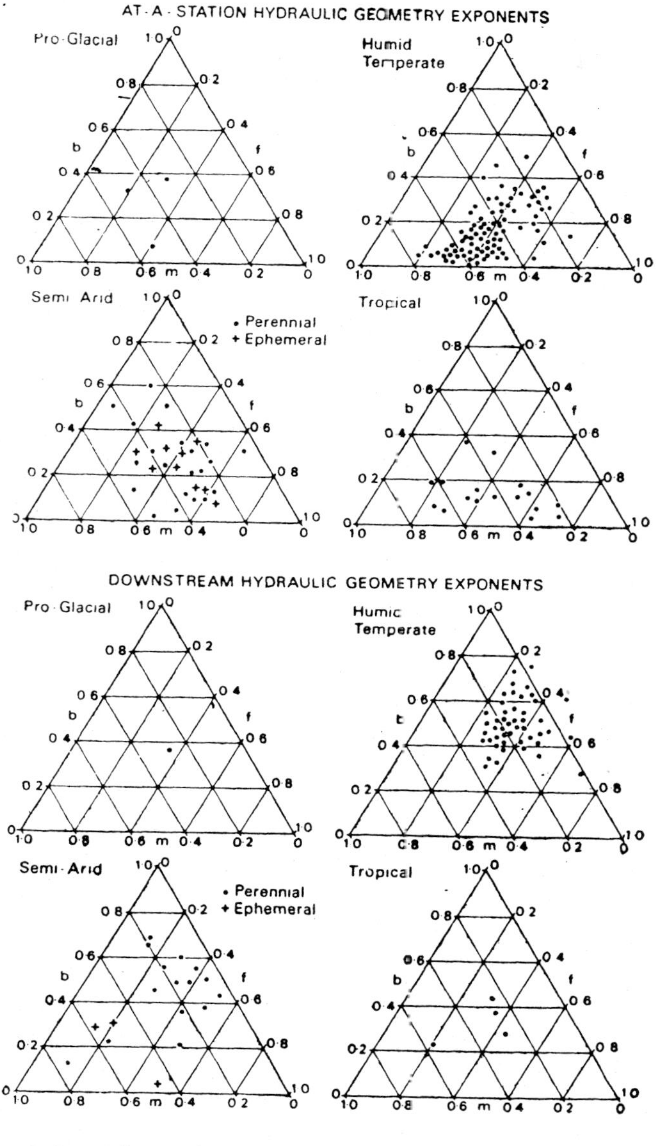

Fig. 2.4 : *Values of hydraulic geometry exponents reported (after Park, 1977).*

of channel enlargement. Scott (1978) has also identified some esoteric aspects of channel geometry in Arctic areas. He concluded that in general permafrost tends to increase bank stability, and that the importance of the spring break-up flood is dependent upon the size of the basin, being more significant in larger basins. This scale dependency helps to explain some earlier contradictory findings.

Differences between downstream changes of channel geometry are also likely to occur as a result of the influence of climate, since the rate of runoff generation from valley-side slopes and evaporative losses from river channels is a direct consequence of the balance between precipitation and evapotranspiration. In the same way that the rate of change of width and depth with discharge at-a-section can be described by hydraulic geometry exponents, the rate of change of these variables in the downstream direction with (normally) increasing discharge can also be described by exponents. Again an analysis of the available data with respect to climate was presented by Park (1977). Humid temperate streams (Fig. 2.4) tend to have medium to high width exponents, medium depth exponents and very low velocity exponents, in comparison with tropical streams which generally have higher velocity exponents. Perennial streams in semi-arid areas have values similar to those of temperate areas, but ephemeral streams in these areas tend to have lower width exponents and relatively high velocity exponents. This Schumm (1977) ascribes to the high water loss in ephemeral stream channels and to the increase in suspended sediment concentrations in a downstream direction.

Channel geometry variations do not exist independently of channel pattern variations, and Langbein and Leopold (1966) have linked variation of channel plan pattern and bed profile in their minimum variance theory of channel adjustment to environmental characteristics. Schumm (1977) has also provided a series of descriptive equations to describe how changes of sediment or water discharge may affect stream cross-sectional geometry and channel pattern. Although intended for application to temporal changes of process rates the relationships embodied in the equations are also relevant in the spatial sense within a restricted area. Four combinations of changing discharge and sediment load exist. Plus and minus exponents indicate an increase or decrease in a variable, and the resulting equations are :

$$Q^+ \; Qs^+ \sim b^+, \; d^\pm, \; \lambda^+, \; S^\pm, \; P^-, \; F^+$$
$$Q^- \; Qs^- \sim b^-, \; d^\pm, \; \lambda^-, \; S^\pm, \; P^+, \; F^-$$
$$Q^+ \; Qs^- \sim b^\pm, \; d^+, \; \lambda^-, \; S^-, \; P^+, \; F^-$$
$$Q^- \; Qs^+ \sim b^\pm, \; d^-, \; \lambda^-, \; S^+, \; P^-, \; F^+$$

where the symbols are :

Q — water discharge
Qs — bed load sediment discharge
b — channel width
d — channel depth
λ — meander wavelength
S — gradient
P — channel sinuosity
F — width/depth ratio

Thus, from the first equation an increase in water and bed-load discharge results in an increase in width, meander wavelength and width/depth ratio, and a decrease in sinuosity. Channel depth is affected by changes in water and bedload sediment discharge in opposite senses, and the direction of the resulting change is, therefore, indeterminate, as is also the case for gradient. The total change within the system is distributed amongst the variables in such a way as to minimise the total variance of the variables, according to the minimum variance theory as illustrated by Williams (1978). Other combinations of changes in water and bedload discharge may be evaluated in a similar fashion from the other equations. Although this model is expressed in terms of discharge of water and sediment these are closely related to climate, since not only does water discharge reflect the balance between rainfall and evapotranspiration, but sediment discharge also reflects the overall climatic characteristics of the area. This is because the sediment load is largely controlled by vegetation characteristics in the basin, which are in turn dependent upon climate.

Much palaeohydrological interpretation of fluvial sediments has been based on such relationships between morphological characteristics of palaeochannels and former discharges, and changed discharges have been related to changed climatic conditions. For example, Schumm (1968) related three different types of channel in the deposits of the Riverine Plains of Australia to three different climatic regimes, of which today's is intermediate in character between a more arid climate, responsible for a wider and shallower palaeo-channel, of lower sinuosity than the present-day channel; and a more humid climate, responsible for a wider and deeper channel of increased sinuosity when compared with today's channel.

Aspects of channel pattern and cross-sectional geometry are intimately related to one another and to discharge of both water and sediment, which are in turn related to climate. It is, therefore, both difficult and conceptually inadvisable to attempt to identify the impact of climate on single aspects of the fluvial system, since variation is distributed amongst all of the components of the system. Thus, Gregory (1977a), in attempting to explain changes of channel capacity in New South Wales, was forced to examine these within the context of the whole network, and to develop indices facilitating this (Gregory, 1977b, 1979), as used by Park (1979). This problem is exacerbated by the existence of what Schumm (1977) has termed Complex Response within fluvial systems, whereby the nature and sequence of the adjustments experienced depend not only upon the nature of the changes in environmental characteristics experienced but also upon spatial location within the basin. It is, therefore, necessary to examine river channels within the context of the networks in which they are embedded.

At the highest level of fluvial organisation, the drainage basin or network, climate again exerts an influence on fluvial characteristics. The impact of climate at this scale has been extensively examined by Gregory (1976), who examined much of the available data. Drainage density is the morphometric parameter which has been most fully examined. For example, Gregory and Gardiner (1975) suggested that drainage density can be approximately related to mean annual precipitation, when adjusted for variations in basin size (Fig. 2.5). The highest values are found in semi-arid areas (e.g. Melton, 1957; Woodyer and Brookfield, 1966) or in areas with markedly seasonal precipitation regimes (e.g. Smith, 1950; Maxwell, 1960;

Brice, 1966). In humid mid-latitude areas drainage densities are generally low, with an overall tendency for increasing values with increasing precipitation. Humid tropical regions may be characterised by relatively high values, as suggested earlier. In addition to this climatic variation of mean values of drainage density, the range of values reported also varies according to climate, in a pattern similar to that of the mean values. The highest range occurs in semi-arid areas, and low ranges in humid mid-latitudes. This pattern is also analogous to that identified for sediment yield in relation to mean annual precipitation (Schumm, 1965; Wilson, 1973).

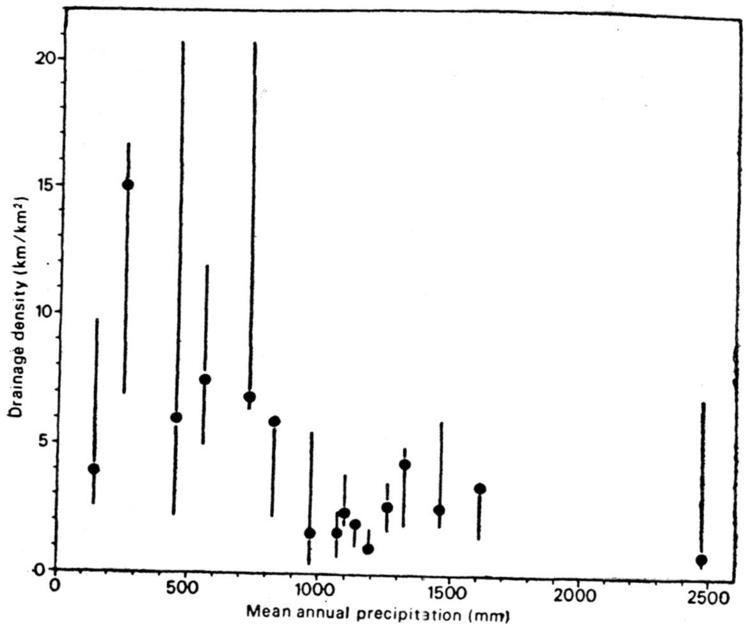

Fig. 2.5 : *Relationship between drainage density and mean annual precipitation*

Drainage density values are obtained from regression relationships between basin area and total stream length, with the circle representing the drainage density indicated by the regression relationship for a basin of one square kilometre, and the black line representing the range of values reported. (After Gregory and Gardiner, 1975, and Gregory, 1976, and with additional data from the Belgian basin examined.

Although variation in mean annual precipitation explains some of the variation in reported values of drainage density, indices of other aspects of climate have also been employed, as well as indices of climate in combination with other environmental attributes, especially vegetation. For example, Gregory (1976) examined indices of rainfall intensity and the radiation index of dryness, Melton (1957) employed the Thornthwaite Potential Evapotranspiration index and rainfall intensity, and Chorley (1958) developed a vegetation index incorporating precipitation effectiveness, amount and intensity.

Whilst climate undoubtedly exerts an influence on drainage density at the global scale studies at more local scales suggest that other factors are also important. Although climate, and particularly rainfall intensity, is suggested to be important at the regional scale by studies such as those of Karsten and Tuttle (1971) and Gregory and Gardiner (1975) the nature of the control may differ in different regions. For example, Madduma Bandara (1974) suggested that above a certain threshold level of precipitation drainage density increases with increasing values of the Thornthwaite Potential Evapotranspiration index, whereas Melton (1957) had suggested the reverse (inverse) relationship for a less humid area. Other studies (e.g. Wilson, 1971; Gardiner, 1971, 1974) also suggest that rock type assumes an increasing significance as the spatial scale is reduced.

Although much information is now available concerning the variation of drainage density with climate very little attention has been directed towards the impact of climate upon other aspects of river networks, and Gardiner (1978) suggested that whereas density and slope aspects of the basin may depend upon environmental characteristics, including cilmate, other morphometric components such as basin shape and network geometry may depend to a greater extent on space-filling properties of the network. They are, therefore, largely conditioned by the random model of network composition and are independent of environmental characteristics, assuming that no marked geological inhomogeneities or anisotropies exist. However, research continues into the development of methods by which these characteristics of the network may be more effectively measured and related to environmental characteristics (e.g. Jarvis, 1976), and relatively subtle influences of climate upon network topology and drainage composition may yet be identified.

In conclusion, it can be seen that climate undoubtedly exerts an influence on fluvial networks at all levels of organisation. Study of this influence has proceeded from the empirical identification of largely 'black-box' relationships at the basin scale in the 1950s, towards investigation of smaller-scale relationships in which processes and theoretical modelling are involved. This has resulted from the necessity to understand the processes responsible for relationships within the fluvial system. The successful integration between empirically-identified relationships and process-related explanations is as yet incomplete, and is hindered by an incomplete, and spatially uneven availability of data and observations. Sufficient results are, however, available to suggest that this field may be one of considerable geomorphological value.

ACKNOWLEDGMENTS

Fieldwork during which some of the data presented in Figure 2.5 and Table 2.2 were gathered was financed by a European Faculty Fellowship from the Leverhulme Trust.

SUGGESTED READINGS

Abrahams, A.D., 1980 : Channel link density and ground slope, *Annals, Association of American Geographers*, 70, 80-93.

Ballantyne, C.K., 1978 : The hydrologic significance of nivation features in permafrost areas, *Geografiska Annaler*, 60A, 51-4.

Brice, J.C., 1966 : Erosion and deposition in the loess-mantled Great Plains, Medicine Creek drainage basin, Nebraska, *United States Geological Survey Professional Paper*, 352-H, 255-339.

Budel, J., 1963 : Klima-genetische geomorphologie, *Geographische Rundschau*, 7, 269-86., Translated in E. Derbyshire (editor) 1976 *Climatic Geomorphology*.

Chorley, R.J., 1958 : Climate and morphometry, *Journal of Geology* 65, 628-38.

Chorley, R.J., (ed.) 1969 : *Water, Earth and Man*, Methuen : London.

Derbyshire, E., (ed.) 1973 : *Climatic Geomorphology*, Macmillan : London.

Derbyshire, E., (ed.) 1976 : *Geomorphology and Climate*, Wiley : London.

Dunne, T., 1976 : Fieldstudies of hillslope flow processes, in Kirkby, M.J. editor, *Hillslope Hydrology*, Wiley : Chichester.

Dury, G.H., 1969 : Hydraulic Geometry, in Chorley, R.J. editor, *Water, Earth and Man*, Methuen : London.

Gardiner, V., 1971 : A drainage density map of Dartmoor, *Reports and Transactions of the Devonshire Association for the Advancement of Science*, 103, 167-80.

Gardiner, V., 1972 : *A morphometric approach to land form aspects of land classification*, Unpub. Ph.D. thesis, University of Exeter.

Gardiner, V., 1974 : Land form and land classification in North West Devon, *Reports and Transactions of the Devonshire Association for the Advancement of Science*, 106, 141-53.

Gardiner, V., 1978 : Redundancy and spatial organisation of drainage basin form indices, *Transactions of the Institute of British Geographers, New Series*, 3, 416-31.

Gilmour, D.A. and Bonell, M., 1979 : Runoff processes in tropical rainforests with special reference to a study in north-east Australia, in Pitty, A.F. (ed.), *Geographical Approaches to fluvial processes*. Geoabstracts : Norwich.

Gregory, K.J., 1976 : Drainage networks and climate, in Derbyshire, E. editor, *Geomorphology and Climate*, Wiley, London.

Gregory, K.J., 1977a : Channel and network metamorphosis in northern New South Wales, in Gregory, K.J. editor, *River Channel Changes*, Wiley, London.

Gregory, K.J., 1977b : Stream network volume—an index of channel morphometry, *Bulletin of the Geological Society of America*, 88, 1075-80.

Gregory, K.J., 1979 : Drainage network power, *Water Resources Research*, 15, 775-7.

Gregory, K.J. and Gardiner, V., 1975 : Drainage density and climate, *Zeitschrift fur Geomorphologie*, 19, 287-98.

Gregory, K.J. and Gardiner, V., 1979 : Comment on drainage density and streamflow : A closer look, by S.L. Dingman, *Water Resources Research*, 15, 1662-4.

Gregory, K.J. and Walling, D.E., 1973 : *Drainage Basin Form and Process*, Edward Arnold : London.

Gupta, A., 1975 : Stream characteristics in Eastern Jamaica, an environment of seasonal flow and large floods, *American Journal of Science*, 275, 825-47.

Gurnell, A.M., 1978 : The dynamics of a drainage network, *Nordic Hydrology*, 9, 293-306.

Jarvis, R.S., 1976 : Classification of nested tributary basins in analysis of drainage basin shape, *Water Resources Research*, 12, 1151-64.

Karsten, R.A. and Tuttle, S.D., 1971 : The distribution of climatic factors in Iowa and their influence on geomorphic processes, *Iowa Academy of Science Proceedings*, 77, 266-81.

Kennedy, B.A., 1972 : Bankfull discharge and meander forms, *Area* 4, 209-12.

Kirkby, M.J., 1976 : Implications for sediment transport, in Kirkby, M.J. editor, *Hillslope Hydrology*, Chichester : Wiley.

Langbein, W.B. and Leopold, L.B., 1966 : River meanders—theory of minimum variance, *United States Geological Survey Professional Paper*, 422-H.

Leopold, L.B. and Maddock, T., 1953 : The hydraulic geometry of stream channels and some physiographic implications, *United States Geological Survey Professional Paper*, 252.

Löffler, E., 1977 : *Geomorphology of Papua New Guinea*, Australian National University Press : Canberra.

Madduma Bandara, C.M., 1974 : Drainage density and effective precipitation, *Journal of Hydrology*, 21, 187-90.

Maxwell, J.C., 1960 : Quantitative geomorphology of the San Dimas experimental forest, California, *Office of Naval Research, Geography Branch, Project NR 389-042, Technical Report 19.*

McCann, S.B. and Cogley, J.G., 1973 : The geomorphic significance of fluvial activity at high latitudes, in Fahey, B.D. and Thompson, R.D. (eds.), *Research in Polar and Alpine Geomorphology*, Geoabstracts : Norwich.

Melton, M.A., 1957 : An analysis of the relations among elements of climate, surface properties and geomorphology, *Office of Naval Research, Geography Branch, Project NR 389-042, Technical Report, 11.*

Morgan, R.P.C., 1972 : Observations on factors affecting the behaviour of a first-order stream, *Transactions, Institute of British Geographers*, 56, 171-86.

Morgan, R.P.C., 1973 : The influence of scale in climatic geomorphology : A case study of drainage density in West Malaysia, *Geografiska Annaler*, 55A, 107-15.

Morgan, R.P.C., 1976 : The role of climate in the denudation system : A case study from West Malaysia, in Derbyshire, E. (ed.), *Geomorphology and Climate*, Wiley : London.

Pardé, M., 1933 : *Fleuves et Rivières* (1st edition), Armand Colin : Paris.

Park, C.C., 1977 : World-wide variations in hydraulic geometry exponents of stream channels : An analysis and some observations, *Journal of Hydrology*, 33, 133-46.

Park, C.C., 1979 : Tin streaming and channel changes—some preliminary observations from Dartmoor, England, *Catena* 6, 235-44.

Peltier, L.C., 1950 : The geographic cycle in periglacial regions as it is related to climatic geomorphology, *Annals, Association of American Geographers*, 40, 214-36.

Rhodes, D.D., 1977 : The b-f-m diagram : graphical representation and interpretation of at-a-station hydraulic geometry, *American Journal of Science*, 277, 73-96.

Ruxton, B.P., 1967 : Slope wash under mature primary rainforest in Northern Papua, in Jennings, J.N. and Mabbutt, J.A. (eds.), *Landform Studies from Australia and New Guinea.*

Schumm, S.A., 1965 : Quaternary Palaeohydrology, in Wright, H.E. and Frey, D.G. (eds.), *Quaternary of the United States*, Princeton University Press : Princeton.

Schumm, S.A., 1968 : River adjustment to altered hydrologica regimen, Murrumbidgee River and palaeochannels, Australia, *United States Geological Survey Professional Paper 598.*

Schumm, S.A., 1977 : *The Fluvial System*, Wiley : New York.

Scott, K.M., 1978 : Effects of permafrost on stream channel behaviour in Arctic Alaska, *United States Geological Survey Professional Paper 1068.*

Smith, D.G., 1979 : Effects of channel enlargement by river ice processes on bankfull discharge in Alberta, Canada, *Water Resources Research*, 15, 469-75.

Smith, K.G., 1950 : Standards for grading texture of erosional topography, *American Journal of Science*, 248, 655-68.

Stoddart, D.R., 1969 : Climatic geomorphology : Review and assessment, *Progress in Geography*, 7, 160-222.

Ward, R.C., 1975 : *Principles of Hydrology* (2nd edition), McGraw-Hill : London.

Wilkinson, J.J. and Bunting, B.T., 1975 : Overland transport of sediment by rill water in a periglacial environment in the Canadian High Arctic, *Geografiska Annaler*, 57A, 105-16.

Williams, G.P., 1978 : Hydraulic geometry of river cross-sections—theory of minimum variance, *United States Geological Survey Professional Paper, 1029.*

Wilson, L., 1971 : Drainage density, length ratios and lithology in a glaciated area of south Connecticut, *Bulletin, Geological Society of America*, 82, 2955-6.

Wilson, L., 1973 : Relationships between geomorphic processes and modern climates as a method in paleoclimatology, in Derbyshire, E. (ed.), *Climatic Geomorphology*, Macmillan : London.

Woodyer, K.D. and Brookfield, M., 1966 : The land system and its stream net, *C.S.I.R.O. Division of Land Resources, Technical Memo*, 66/5.

THREE

Nadira Kabir and *John F. Orsborn*

RIVER CHANNEL AND BASIN CHARACTERISTICS*

INTRODUCTION

A RIVER channel adjusts its size to accommodate the channel forming flows. The adjustments of the stream channel are reflected in its morphology, and these changes in width, depth, and velocity in response to changes in flow have been observed to have certain characteristics. Flows of equal frequency tend to increase in size with drainage area with attendant increases in channel width, depth, and mean velocity as power functions.

The drainage network and the basin morphology are measures of fluvial process operations in the basin. Studies have shown that river flows are related to the topographic and climatic characteristics of a drainage basin which control the amount, time and space distributions of streamflow. The streamflows reflect precipitation variations, and the differences in streamflows between basins will vary due to differences in their surface and subsurface characteristics. Because both channel and basin characteristics are related to streamflow, then within the integrated physical drainage basin system, channel basin characteristics must be interrelated. The concepts of dimensional analysis were introduced into geomorphology by Strahler with the development of qualitative relationships between various geomorphic factors of a drainage basin and a flow term.

The objective of the study reported herein was to relate channel and basin characteristics using dimensional analysis. Several basin characteristics which might influence the stream-

*Based on reference 6.

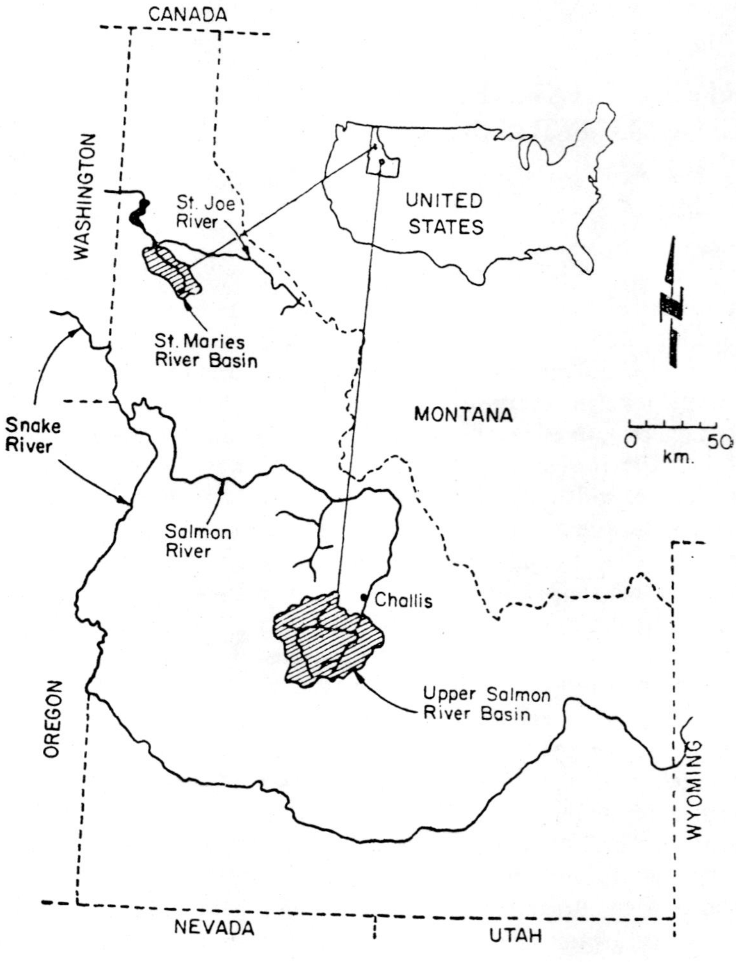

FIG. 3.1 : *Location map of the two study areas in Idaha*

flow and, hence, channel characteristics, were analyzed by application of the Buckingham Pi theorem with the intention of developing generalized expressions applicable to different hydrologic provinces.

Relations between channel and basin characteristics were studied separately for two different areas with diverse river systems. The areas are the Upper Salmon River Basin in south-central Idaho, and the St. Maries River Basin in northern Idaho, as shown in Fig. 3.1. The main features of the study areas are summarized in Table 3.1.

TABLE 3.1 : **Main features of study areas**

Features	Upper Salmon River Basin*	St. Maries River Basin**
Total drainage area, km^2	4,662	1,295
Main stream	Salmon River. . . E. Fork Salmon River	St. Maries River
Headwaters	White Cloud Mtns.	Clearwater Mtns.
Upper elevation, km	19,000	9,660
Outlet elevation, km	8,310	3,220
Precipitation range	25 to 127 cm/yr	76 to 130 cm/yr
Mean precipitation	76 cm/yr	89 cm/yr
Average temperature	2°C	9°C
Flood cause	Spring snowmelt	Midwinter rain and/or snow

*Data used from W.W. Emmett. **Data used from M.Y. Tsang

SELECTION OF PARAMETERS

Channel characteristics

A stream channel is formed and maintained by the flows it carries. A reach of a channel transmits varying amounts of water which increases in the downstream direction as a function of increasing drainage area, with the unit areal contribution of flow usually decreasing in the downstream direction. Bankfull flow has a special significance in that it is the limiting flow the channel can retain. By computing the flow size that transports the largest total amount of sediment over a period of years, Wolman and Miller (cited in 3) concluded that the bankfull stage represents the most effective channel forming flow. This dominant flow, QD, is represented by recurrence intervals varying from 1.5 to 4 yrs. For comparison to be made within and between streams, Leopold and Maddock showed that the selection of a flow of constant frequency is extremely important. Bankfull stage approximates an equal frequency flow and Bankfull channel characteristics are related to basin characteristics in this study. Average values of bankfull width W, depth D, and velocity V, are channel characteristics in the continuity equation. Bankfull channel width W, is defined as the horizontal distance from the top of the lower bank to the opposite bank wall. Channel cross-sectional area A, is defined as channel cross-section which is bounded by the width and the channel wetted perimeter. The mean bankfull depth D, is defined as the quotient of the cross-sectional area and the bankfull width. Mean velocity V, is computed from $(QD)/A$.

Basin characteristics

Conceptually, the physical characteristics of a basin which might influence channel forming flows usually need to be expressed by representative indexes as given below.

Size of the basin area AB, can be considered a logical cause of streamflow variation between sites. Area of the basin represents the catchment size for precipitation and, hence, governs the average amount of flow of a stream.

Basin relief HB, is the elevation difference between the headwaters and outlet, and is a measure of the potential energy or driving force of a drainage system due to gravity acting on streams and ground water. Basin relief was determined as the elevation difference between the basin outlet and the longest continuous contour above the end of the blue line streams on the U.S. Geological Survey (USGS) topographic maps.

The slope of a drainage basin is another factor which influences streamflow. Direct runoff decreases with decreased steepness of the valley and side slopes. For this study a simple index of slope used by Benson was selected. This index is the average slope of the main stream between points 10 and 85 per cent of the distance upstream from the outlet to the basin border. The symbol *SM* is used for main stream slope.

Basin length *LB*, scaled along the major axis, affects the concentration of the runoff. Basin relief, when divided by basin length, gives the dimensionless relief ratio *RH*. Relief ratio measures the overall steepness of the drainage basin. Because main stream slope and relief ratio are both measures of vertical to horizontal distances, only one may be used as an indicator of the slope of drainage network geometry.

The drainage network is indicative of the fluvial processes operating in the basin. The total stream length is related to the percentage of the precipitation which becomes streamflow and, thus, represents an analogy to the soil type, geology, and basin-storage contribution to low flow. On USGS topographic maps, perennial streams are shown by solid blue lines and intermittent streams by dashed blue lines. In this study, total stream length within a basin *LS*, is defined as the length of the solid and dashed blue lines. Since bankfull flow is used in this study, total stream length in high flow use should include the intermittent streams.

In general, bed material size decreases in the downstream direction and is related to the change of width and depth in the downstream direction. The median size of the stream bed materials, *D50*, was used as the representative grain size *DG*.

Mean annual precipitation *P*, is a measure of the amount of water supplied to a drainage basin, and of the potential runoff. Mean annual basin precipitation was computed for each sub-basin from a U.S. Weather Service precipitation map for Idaho using the isohyetal method.

DIMENSIONAL ANALYSIS

The fundamental dimensions of mass, length and time, whether used singly or combined as products, are sufficient to define all geometrical and mechanical properties of a drainage basin. Combination of dimensional elements produces dimensionless ratios which provide descriptive indexes of terrain, irrespective of scale.

Considering the drainage basin characteristics which might influence the size of streamflow and, thus, channel size and velocity, the following functional relationships can be obtained:

$$W = f_1[AB, HB, SM, DG, LS, QD, \mu, \rho, g] \quad \ldots\ldots (1)$$
$$D = f_2[AB, HB, SM, DG, LS, QD, \mu, \rho, g] \quad \ldots\ldots (2)$$
$$V = f_3[AB, HB, SM, DG, LS, QD, \mu\ \rho, g] \quad \ldots\ldots (3)$$

Where μ=viscosity of water; ρ=density of water; and g=acceleration due to gravity with other terms as previously defined. Dimensions of all the terms used in these expressions are shown in Table 3.2. Use of QD, HB and ρ as repeating variables reduces Eq. 1 to seven dimensionless parameters:

$$\frac{W}{HB} = f_4\left[\frac{AB}{(HB)^2}, SM, \frac{LS}{HB}, \frac{\Gamma G}{HB}, \frac{(QD)\rho}{(HB)\mu}, \frac{(QD)^2}{(HB)^5 g}\right] \ldots (4)$$

TABLE 3.2 : **Dimensions of variables used in the dimensional analysis**

Symbol	Term	Dimension	Symbol
W	Bankfull channel width	Length	L
D	Bankfull mean depth	Length	L
V	Bankfull flow velocity	Distance per unit time	LT^{-1}
AB	Drainage basin area	Length times length	L^2
HB	Basin relief	Length	L
SM	Main stream slope index	Length divided by length	None
RH	Relief ratio	Length divided by length	None
DG	Diameter of the characteristic channel bed material *D50*	Length	L
LS	Total stream length	Length	L
QD	Bankfull or dominant flow	Volume rate of flow	$L^3\ T^{-1}$
μ	Viscosity of fluid	Measure of resistance to flow	$ML^{-1}\ T^{-1}$
ρ	Density of fluid	Mass per unit volume	ML^{-3}
g	Acceleration due to gravity	Distance per unit time, per unit time	LT^{-2}

From the continuity equation, QD is a measure of velocity, and, thus, the inertia of the flow. Therefore, the non-dimensional parameter, $(QD)\rho/(HB)\mu$, includes measures of inertia and viscous forces. For bankfull streamflow, viscous effects in the Reynolds number are small and can be neglected. The basin relief HB, is a measure of the driving force due to gravity acting on all water in a basin. Thus, the non-dimensional parameter $(QD)^2/(HB)^5g$, represents a ratio of inertia to gravity forces and is a geomorphic-hydraulic basin Froude number F_B. Considering the above reasoning, Eq. 4 can be written as :

$$\frac{W}{HB}=f_5\left[\frac{AB}{(HB)^2}, SM, \frac{LS}{HB}, \frac{DG}{HB}, F_B\right] \quad \ldots\ldots (5)$$

The non-dimensional ratios were combined into a group by multiplication and division so that the results are physically consistent with the observed conditions. Proceeding downstream, the discharge tends to increase, whereas the stream density tends to decrease with increasing drainage area. With width, depth, and velocity increasing in the downstream direction, discharge and drainage area also increase, while channel slope and bed material size decrease. So, Eq. 5 may be written as :

$$\frac{W}{BH}=f_6\left[\frac{1}{SM}\cdot\frac{AB}{(LS)\ (DG)}.F_B\right]=f_6\,[X_1] \quad \ldots\ldots (6)$$

in which $X_1 = [1/SM].\ [AB/(LS)\ (DG)]\ .\ F_B$. Similarly, applying the same analysis and arguments, it can be shown that :

$$\frac{D}{HB}=f_7\left[\frac{1}{SM}\cdot\frac{AB}{(LS)\ (DG)}.F_B\right]=f_7\,[X_1] \quad \ldots\ldots (7)$$

and

$$\frac{V(HB)^2}{QD}=f_8\left[\frac{1}{SM}\cdot\frac{AB}{(LS)\ (DG)}.F_B\right]=f\,[X_1] \quad \ldots\ldots (8)$$

The above three equations give a set of non-dimensional relationships of bankfull width, mean depth and velocity with main-stream slope, the ratio of basin area to total stream length and characteristic bed material size, and the basin Froude number. The tests of these equations were conducted using data from two river systems as defined earlier. For the two study areas, four types of data were utilized: (1) the bankfull channel cross-sectional geometries; (2) streamflow values at each station; (3) basin geomorphic parameters; and (4) characteristic particle size of bed materials.

RESULTS

Non-dimensional plots were made according to Eqs. 6, 7 and 8 for both study areas and log-transformed linear regression equations were fitted to the plots. Non-dimensional plots to develop correlations for bankfull width, mean depth and velocity for the Upper Salmon River Basin are shown in Fig. 3.2 and for the St. Maries River Basin in Fig. 3.3. The regression lines through the data points of Fig. 3.2 are described by Eqs. 9, 10 and 11 for the Upper Salmon River Basin :

$$\frac{W}{HB}=0.022\left[\frac{1}{SM}\cdot\frac{AB}{(LS)(DG)}.F_B\right]^{0\cdot33}=0.022(X_1)^{0\cdot33} \quad \ldots(9)$$

$$\frac{D}{H_B}=0.00094\,(X_1)^{0\cdot20} \quad \ldots\ldots(10)$$

$$\frac{V\,(HB)^2}{QD}=47{,}470\,(X_1)^{-0\cdot54} \quad \ldots\ldots(11)$$

For the St. Maries River Basin, regression lines through the data points of Fig. 3.3 are described by Eqs. 12, 13 and 14 :

$$\frac{W}{HB}=0.04(X_1)^{0\cdot42} \quad \ldots\ldots(12)$$

$$\frac{D}{HB}=0.00227(X_1)^{0\cdot20} \quad \ldots\ldots(13)$$

$$\frac{V(HB)^2}{QD}=10{,}940(X_1)^{-0\cdot62} \quad \ldots\ldots(14)$$

The main stream slope index *SM*, was used to represent the drainage basin slope. As discussed in a previous section, relief ratio *RH*, also is a measure of drainage basin slope. Non-dimensional plots similar to Figs. 3.2 and 3.3 were made by replacing *SM* in Eqs. 1, 2 and 3 with the relief ratio *RH*, and reducing the functional relationships in the form of Eqs. 6, 7 and 8, designating the set of terms with *RH* as X_2.

The coefficients for all the log-transformed regression

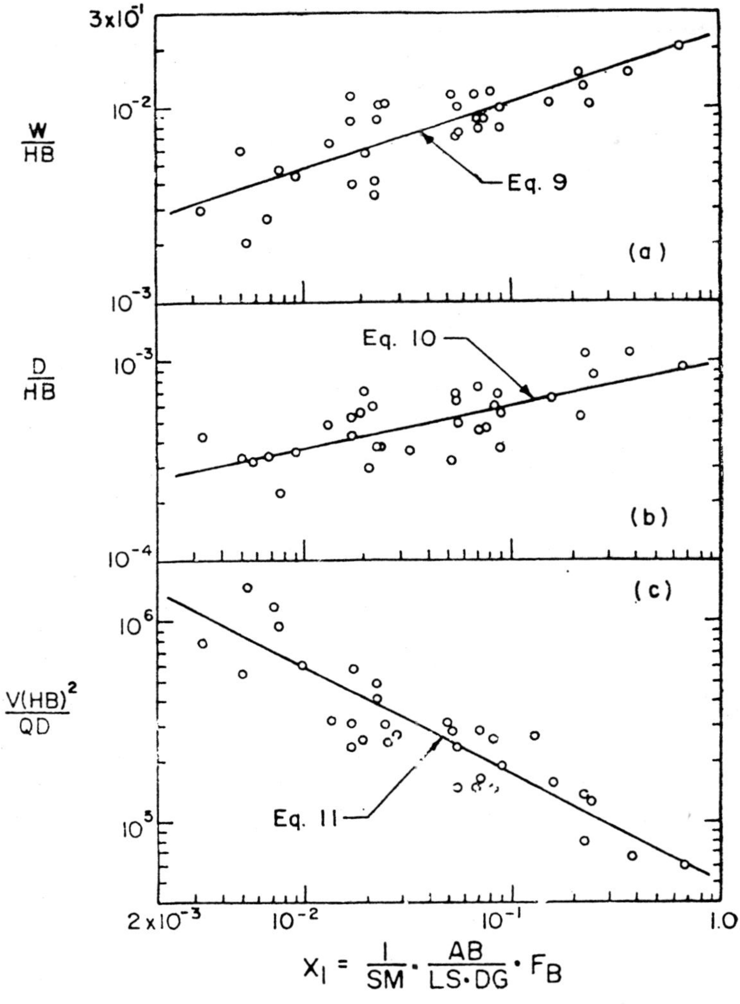

FIG. 3.2 : *Non-dimensional relationships for bankfull width, mean depth, and mean velocity with x_1 for Upper Salman River Basin.*

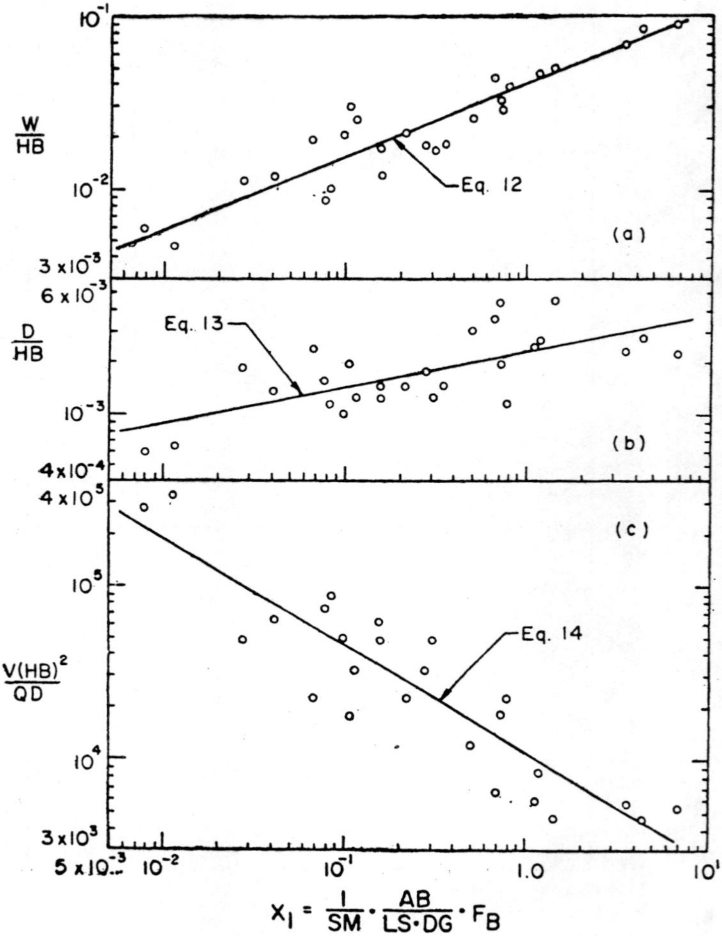

FIG. 3.3 : *Non-dimensional relationships for banffull width, mean depth, and mean velocity with x_1 for St. Maries River Basin.*

equations, together with their correlation coefficients *r*, are given in Table 3.3. The coerrelation coefficient is a measure of linear dependence between two parameters *X* and *Y* with a maximum value of unity. A positive value indicates direct correlation, and a negative value indicates an inverse correlation. The standard errors of estimate of data points for both the basins are given in Table 3.3 and are measures of the scatter of data points about the regression line.

Reduction of results

In the equations resulting from the dimensional analysis for the two study areas the basin Froude number F_B, contains the dominant discharge Q_D. These equations express non-dimensional relations for bankfull width, mean depth and velocity with basin characteristics, dominant discharge and mean diameter of the characteristic bed materials. The dominant discharge can be expressed in terms of basin characteristics and in the study for St. Maries River Basin the two-year peak flood was strongly related to basin characteristics by :

$$QF2P = QD = 1.83\ (P.AB.\sqrt{HB})^{0.85} \qquad \ldots\ldots (15)$$

A similar relationship was developed for the Upper Salmon River Basin (4) described by the following equation :

$$QD = 4.15\ (P.AB.\ \sqrt{HB})^{0.64} \qquad \ldots\ldots (16)$$

Some of the flow data for Eq. 16 were based on very short periods of record, but the regression analysis assumes equal weight for each data point in the set. Otherwise the coefficients and exponents in Eqs. 15 and 16 would be similar.

When Eq. 16 is substituted into Eqs. 9, 10 and 11, the following relationships result for the Upper Salmon River Basin :

$$W = 0.0198 \left[\frac{1}{SM} \cdot \frac{AB^{1.64}}{LS.DG} P^{0.64} . HB^{0.82}\right]^{0.33} \qquad \ldots. (17)$$

$$D = 0.000883 \left[\frac{1}{SM} \cdot \frac{AB^{1.64}}{LS.DG} . P^{0.64} . HB^{2.82}\right]^{0.20} \qquad \ldots (18)$$

$$V = 232{,}928 \left[SM\ . \frac{LS.DG}{AB^{0.46}} \ \frac{P^{0.54}}{HB^{0.93}}\right]^{0.54} \qquad \ldots\ldots (19)$$

Similarly, substituting Eq. 15 into Eqs. 12, 13 and 14 results in the following relationships for St. Maries River Basin :

$$W = 0.0249 \left[\frac{1}{SM} \cdot \frac{AB^{1.85}}{LS.DG} . P^{0.85} . HB^{0.31}\right]^{0.42} \qquad \ldots (20)$$

TABLE 3.3 : **Results of regression analysis**

General Equation $Y=mx^n$		Upper Salmon River Basin				St. Maries River Basin			
Y	X	m	n	r	Standard Error of Estimate	m	n	r	Standard Error of Estimate
$\frac{W}{HB}$	X_1	0.022	0.33	0.80	0.00235	0.04	0.42	0.92	0.00899
$\frac{D}{HB}$	X_1	0.00094	0.20	0.70	0.000149	0.00227	0.20	0.69	0.000755
$\frac{V(HB)^2}{QD}$	X_1	47,470	−0.54	−0.90	138,502	10,940	−0.62	−0.89	34,294
$\frac{W}{HB}$	X_2	0.03	0.35	0.75	0.00261	0.05	0.49	0.89	0.0106
$\frac{D}{HB}$	X_2	0.00119	0.24	0.73	0.000143	0.00257	0.23	0.63	0.000803
$\frac{V(HB)^2}{QD}$	X_2	29,440	−0.59	−0.88	152,385	7,750	−0.74	−0.84	41,295

Note: $X_1 = \frac{1}{SM} \cdot \frac{AB}{LS.DG} \cdot F_B$; $X_2 = \frac{1}{RH} \cdot \frac{AB}{LS.DG} \cdot F_B$.

$$D = 0.00181\left[\frac{1}{SM}\cdot\frac{AB^{1\cdot85}}{LS.DG}\cdot P^{0\cdot85}\,.\,HB^{2\cdot92}\right]^{0\cdot20} \quad \ldots\ldots (21)$$

$$V = 43{,}940\left[SM\cdot\frac{LS.DG}{AB^{0\cdot48}}\cdot\frac{P^{0\cdot52}}{HB^{0\cdot47}}\right]^{0.62} \quad \ldots\ldots (22)$$

Equations 17-22 express relationships between channel and basin characteristics for the two study areas with different river systems.

Discussion of results

The purpose of this study was to determine whether relationships exist between the stream channel characteristics and geomorphic characteristics of a drainage basin. For the two study areas, these relationships have been shown to exist in Eqs. 17-22 which can be written in simpler form. For example, Eq. 20 can be written as :

$$W = 0.025\,\frac{(AB)^{0\cdot78}\ (P)^{0\cdot36}\ (HB)^{0\cdot13}}{(SM)^{0\cdot42}\ (LS)^{0\cdot42}\ (DG)^{0\cdot42}}$$

$$= 0.025\left[\frac{AB}{LS}\right]^{0\cdot42}\frac{(AB.P)^{0\cdot36}\ (HB)^{0\cdot13}}{(SM)^{0\cdot42}\ (DG)^{0\cdot42}} \quad \ldots\ldots (23)$$

From Eq. 23, stream top width varies directly with drainage area, precipitation and relief and inversely with stream density, main-stream slope and characteristic bed material size. Similarly, Eqs. 17-19 and Eqs. 21 and 22 may be written in simplified forms.

The non-dimensional relationships shown in Figs. 3.2 and 3.3 are expressed by log-transformed linear regression equations. The deviations of certain individual points from the average relation are due to the following reasons :

1. Variation between longitudinal profiles of different streams within a homogeneous climatic and physiographic environment may be caused by geologic factors. Studies by Hack and Brush show that channel slopes and particle size of bed materials are dependent upon the type of bedrock materials upon which the stream flows. The bedrock structure affects topography and disposal of precipitation by determining drainage pattern and basin shape.

2. Usually bed material size decreases in the downstream direction whereas width increases along the profile. But for

the two areas studied, there was no such general trend in particle size and width. The reason for this is that both study areas are in mountainous terrain. Tributary actions contribute to the changes in downstream particle size by adding large particles to the main stream, thus, tending to keep particle size fairly large in the main stream.

3. Main stream slope index was used as a measure of basin steepness, but the size of the bed material that can be carried by a river at a section is influenced more by the slope at that section rather than the basin slope.

4. Part of the variability is due to the techniques of data collection. At some locations, for example station 33 in the Upper Salmon River Basin, the width of the stream was confined by a bridge revetment so that bankfull channel width is less than that of the average relationship. Also, there may be some error in determining the bankfull width or discharge, especially for small streams with poorly defined flood plains. For the Upper Salmon River Basin, only a few stations have a length of record of about 10 yrs. Thus, the bankfull or dominant discharges are no doubt erroneous.

In spite of some deviations, average relationships have significant correlation coefficients (Table 3.3), indicating the validity of the relationships. The coefficients and exponents of the non-dimensional relationships are somewhat different for the two river basins. This may be due in part to the difference in geomorphic and climatic conditions between the two basins (Table 3.1 and 3.3). Although both study areas are situated in the headwater regions, the Upper Salmon River Basin is located at higher elevations than the St. Maries River Basin. The lack of similarity in the relationships between the two basins may be the result of dissimilar landforms, but better flow data would most certainly improve the similarity.

CONCLUSIONS

This study used data gathered independently from two separate river systems of Idaho to examine whether or not quantitative relationships exist between channel and basin characteristics. Dimensional analysis was used to express non-dimensional relationships for bankfull width, mean depth and velocity in terms of basin characteristics. Several quantitative geomorphic factors which affect streamflow, and which can be obtained easily from published sources, were used as basin characteristics. The following conclusions can be drawn from the results of this study.

1. Bankfull channel characteristics of width, depth, and velocity of various locations in a river system can be related to their drainage basin characteristics. Because a large number of variables are involved in the formation of a river system, the parametric relationships developed in this study between channel and basin characteristics are of a general form.

2. The non-dimensional groups of basin characteristics were formed such that they were physically consistent with observed conditions. For example, in the downstream direction channel width increases with increasing drainage area, discharge, and relief, whereas channel slope, bed material size, and stream density decrease in the downstream direction, neglecting local variability.

3. The sum of the exponents and the product of the coefficients in the non-dimensional relationships for width, depth, and velocity are at or near unity and, thus, satisfy the continuity equation.

4. The coefficients and exponents of the nondimensional relationships are somewhat different for the two river systems, probably due in part to different geomorphic and climatic conditions, but also due to a lack of streamflow records.

5. The correlation coefficients between channel mean depth and the basin characteristics is the smallest in both river systems. This is because mean depth was estimated by dividing the channel cross-sectional area by top width of the channel. This depth is the depth of an equivalent rectangular section, but natural channel sections usually deviate from a rectangular section and flow and, thus, channel forming force is not distributed uniformly across the section.

6. Use of main stream slope index as a measure of basin slope instead of relief ratio gives better correlation coefficients for almost all the non-dimensional relations. This is logical because as the runoff leaves the drainage area and enters the stream, channel slope becomes the controlling factor rather than the land slope.

7. For both study areas, median particle sizes of bed

materials do not show inverse relationships with channel width as is usually expected when considering changes in the downstream direction. The reason for this may be the tributary action. Tributaries, in general, yield larger particles to the main stream which would tend to keep particle sizes fairly large in the main stream.

In this study, product forms of the nondimensional ratios obtained from dimensional analysis were used. Other combinations, including additive forms of these ratios, may give better correlations. Since the basic relationships have been set by simple correlations, further study is being conducted with multiple correlation to refine the relationships derived through dimensional analysis.

SUGGESTED READINGS

Benson, M.A., 1964 : Factors affecting the occurrence of floods in the south-west, *United States Geological Survey Water Supply Paper 1580-D*, Washington, D.C.

Brush, L.M., Jr., 1961 : Drainage basins, channels and flow characteristics of selected streams in central Pennsylvania, *United States Geological Survey Professional Paper 282-F*.

Dunne, T., and Leopold, L.B., 1978 : *Water in Environmental Planning*, W.H. Freeman and Co., San Francisco, Calif.

Emmett, W.W., 1975 : The channels and waters of the upper Salmon river area, Idaho, *United States Geological Survey Professional Paper 870-A*.

Hack, J.T., 1957 : Studies of longitudinal stream profiles in Virginia Maryland, *United States Geological Survey Professional Paper* and *294-B*.

Kabir, N., 1980 : River characteristics from basin characteristics, M.S. Thesis, Dept. of Civil and Environmental Engineering, Washington State University, Pullman,

Langbein, W.B. et al., 1947 : Topographic characteristics of drainage basins, *United States Geological Survey Water Supply Paper 968-C*.

Leopold, L.B. and Maddock, T., Jr., 1953 : The hydraulic geometry of stream channels and some physiographic implications, *United States Geological Survey Professional Paper 252*.

Leopold, L.B.; Wolman, M.G., and Miller, J.P., 1964 : *Fluvial Processes in Geomorphology*, W.H. Freeman and Co., San Francisco, Calif.

Morisawa, M.E., 1968 : *Streams, their Dynamics and Morphology*, McGraw-Hill Book Co., New York, N.Y.

Orsborn, J.F., Dec., 1974 : Determining streamflows from geomorphic parameters, *Journal of the Irrigation and Drainage Division*, ASCE, Vol. 100, No. IR4, Proc. Paper 10986, pp. 455-475.

Strahler, A.N., Mar., 1958 : Dimensional analysis applied to fluvially eroded landforms, *Geological Society of America Bulletin*, Vol. 69, pp. 279-299.

Thomas, D.M., and Benson, M.A., 1970 : Generalization of streamflow characteristics from drainage basin characteristics, *United States Geological Survey Water Supply Paper 1975*, Washington, D.C.

Tsang, M.Y., 1980 : *Relationships among Streamflow, Drainage Basin, and Channel Characteristics*, M.S. Thesis, Dept. of Civil and Environmental Engineering, Washington State University, Pullman.

FOUR

Athol D. Abrahams and *Alan J. Miller*

AN EMPIRICAL APPROACH TO THE FREQUENCY DISTRIBUTION OF EXTERIOR LINK-LENGTHS IN NATURAL CHANNEL NETWORKS

INTRODUCTION

EXTERIOR channel links or first-order Strahler streams are sections of channel extending without intervening forks from a source at their upstream end to a fork at their downstream end (Shreve 1966, p. 20; 1967, p. 178). Early investigations of exterior link-lengths in the United States (Miller 1953, Table 2 (1); Strahler 1954, p. 7; Schumm 1956, p. 607-608; Maxwell 1960, p. 17) and a more recent study in England (Gardiner 1973, p. 149, Table 2) have indicated that the frequency distributions of such lengths measured on topographic maps can be closely approximated by the log-normal probability density.

However, in computer-simulated channel networks generated according to the random walk model of Leopold and Langbein (1962, p. 18), Smart et al. (1967, p. 97), found that exterior link-lengths conform to the geometric probability density, which is equivalent to the exponential density for a continuous length variable. Although the exponential density for exterior link-lengths has the advantage of possessing a simple underlying rationale (Smart et al. 1967, p. 97; Shreve 1969, p. 404), this density is difficult to reconcile with the above mentioned empirical work pointing to the log-normal density. A possible explanation for this discrepancy has been offered by Kirkby (see Shreve 1969, p. 400) who suggested "that links measured in the field fit an exponential distribution

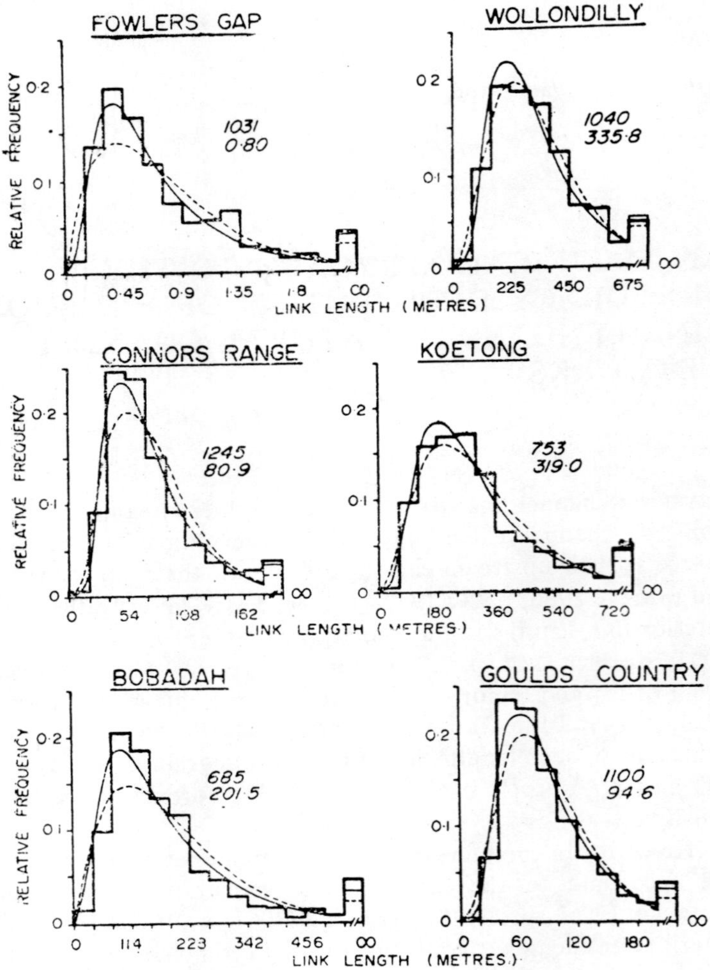

FIG. 4.1 : *Relative frequent distribution of exterior link-lengths.*

The six Australian study areas together with the fitted log-normal (solid line) and gamma (dashed line) distributions. The class interval is equal to one quarter of the medium exterior link-length. The numbers besides each distribution indicate the sample size and mean link-length in metres.

fairly well, but that those measured on maps fit a distribution more like the log normal." This explanation, however, is not supported by the exterior link-length data obtained by field measurement of 73 miniature channel networks at Fowlers Gap (described below). As Fig. 4.1 shows, the exterior link-length distribution for these networks has relatively few short links, and bears no resemblance to the exponential density. When this result is considered along with the results of the map studies cited above, the exponential density emerges as a poor model for exterior link-lengths in natural channel networks, however, they are measured.

A third model for exterior link-lengths has been introduced by Krumbein and Shreve (1970, p. 82, 84-87) who fitted gamma distributions to these lengths in magnitude 5 and magnitude 10 networks in Eastern Kentucky. Dunkerley (1977, p. 462) also fitted gamma distributions to exterior link-lengths in channel networks developed on emergent coral reef terraces in Papua New Guinea. Unlike Krumbein and Shreve, Dunkerley tested the goodness of fit of the gamma distributions, and reported that it was satisfactory in most, but not all cases. Although the gamma is a natural generalization of the exponential density (Smart 1972, p. 323), the gamma density for link-lengths has no underlying rationale (Shreve 1969, p. 403). In other words, the gamma density, like the log normal, appears to have been selected mainly because it provides a reasonably good fit to the data. However, a mechanism which gives rise to gamma-like distributions of exterior link-lengths has been recently suggested by Dunkerley (1977, pp. 465-469). Dunkerley performed a computer-simulation experiment in which channel networks expanded headward and bifurcated in such a manner as to maintain a certain drainage density. He found that exterior links in these networks had length distributions that could be acceptably fitted by the gamma density, and concluded that "the gamma-like density may be inherent in a (drainage) density-controlled branching process" (p. 468).

Thus, two apparently satisfactory models for exterior link-lengths have been proposed in the literature: the log-normal density and the gamma density. However, from the evidence available, it is not possible to determine whether these two densities are merely good empirical approximations to at least some sets of exterior link-length data, or whether exterior link-lengths do in fact conform to one of these models. Any investigation of this question is complicated by the fact that there exist two types of exterior link termed source (*S*) links (exterior links which join another exterior link downstream) and tributary-source (*TS*) links (exterior links which join as interior link downstream) (Mock 1971, p. 1558), and these two types of link have demonstrably different length distributions (Abrahams and Campbell 1976). This observation has important implications for the distributions of exterior link-lengths for each such distribution is a composite or mixture of an *S* and *TS* link-length distribution. Thus, for example, if *S* and *TS* link-lengths alike have log-normal distributions with different means and/or standard deviations then the composite exterior link-length distribution cannot be log-normal, though there may be a log-normal distribution which is a close approximation. A similar comment applies if *S* and *TS* link-lengths have gamma distributions.

The objects of this study are, therefore, (1) to compare the log-normal and gamma densities as empirical approximations to the distributions of exterior, *S* and *TS* link-lengths and (2) to ascertain whether any of these types of link-lengths actually conform to either of these two densities. The study is based on 12 sets of link-length data collected from 11 areas representing a variety of relatively uniform environments, reliefs and stages of landscape development. Owing to the scarcity of reliable data for exterior link-lengths, the data analyzed here are of necessity rather variable in quality. Nevertheless, remarkably consistent results are obtained from all 11 areas. The nature of the errors in the data and their effects on the observed link-length distributions are discussed in a later part of the paper.

Study areas and data

The climate, geology, relief and stage of landscape development in each study area are briefly summarized in Table 4.1. These data give an indication of the broad range of environmental and geomorphic conditions that are present in the 11 areas. Additional information on these conditions is contained in the references accompanying the following descriptions of the link-length data.

Fowlers Gap, Wollondilly, Connors Range, Koetong, Bobadah and Goulds Country

These areas are all located in eastern Australia, and have been recently investigated by Abrahams and Campbell (1976). At Fowlers Gap, a large number of miniature channel networks have developed in an alluvial and colluvial valley fill. Seventy-three of these networks, ranging in magnitude from 5 to 56, were mapped in the field, and their exterior link-lengths measured along the ground. In the Wollondilly area, 31 networks with magnitudes between 5 and 177 were mapped on 1:15,840 scale topographic maps with a contour interval of 25 ft. Subsequent field checking has revealed that valley heads and channel heads in this area do not always coincide, and that the mapped networks represent actual valley networks rather than channel networks. Because the exterior link-lengths were measured directly on the topographic maps, the recorded lengths represent the true lengths somewhat shortened by projection onto the horizontal plane. In each of the Connors Range, Koetong, Bobadah and Goulds country areas, a large valley network was mapped from air photographs and checked to a limited extent in the field. The exterior link-lengths in these areas were measured either directly on the air photographs or on enlarged planimetric maps prepared from the photographs. Relative frequency distributions for the lengths of the exterior links in these six areas are depicted in Fig. 4.1. Similar distributions for the lengths of S and TS links in these areas have been published by Abrahams and Campbell (1976, Figs. 2 and 3).

Perth Amboy

Schumm (1956) carried out a comprehensive study of a magnitude 214 channel network in a small area of badlands developed on a terrace-like, clay-sand fill at Perth Amboy, New Jersey. The lengths of the exterior links in this network were remeasured on the map published by Schumm (1956, Pl, 1, facing p. 597) using a Hewlett-Packard 9864A digitizer.

TABLE 4.1 : **Characteristics of study areas**

Area	Climate*	Geology	Drainage Relief** (metres)	Stage of Landscape Development
(1)	(2)	(3)	(4)	(5)
Fowlers Gap	BWh	Unconsolidated sediments	0.25—0.75	Channel networks expanding headward into flat upper surface
Wollondilly	Cfa	Acid lavas, pyroclastics and porphyrites	180—360	Rejuvenated but little upland surface remaining
Connors Range	Cfa	Granite	75—200	Mature but nickpoints evident in headwaters
Koetong	Cfb	Adamellite	60—200	Mature but nickpoints evident in lower part of basin
Bobadh	BSh	Granite	10—40	Mature
Goulds Country	Cfb	Adamellite	10—30	Mature
Perth Amboy	Cfa	Unconsolidated sediments	3—8	Channel network expanding headward into flat upper surface
Eastern Kentucky	Cfa	Flat-lying sand-stone with inter-bedded siltstone, shale and coal	90—120	Mature

Blountville	Cfa	Tightly folded sandstone and thinly bedded shale	75—100	Mature
Pennington	Cfa	Gently dipping interbedded shale and sandstone	150—200	Mature
Copper Ridge	Cfa	Gently dipping crystalline dolomite	90—120	Mature

*Koppen classification: BWh=hot desert; BSh=hot steppe; Cfa=humid temperate (warmest month mean>22°C); Cfb=humid temperate (warmest month mean <22°C).

**Drainage relief is defined as the vertical distance between a stream and the adjacent divides (see Mark 1975, p. 169).

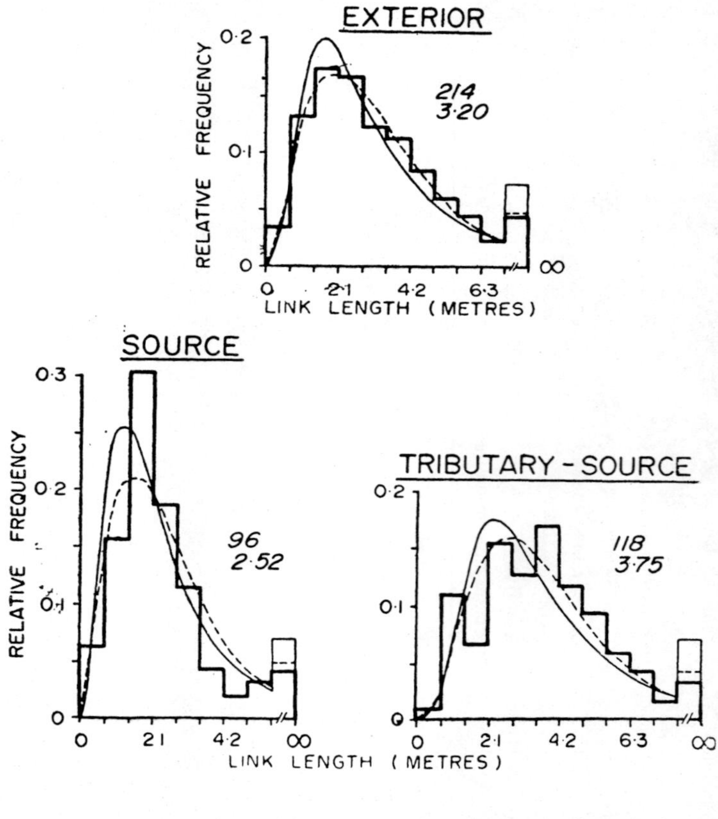

FIG. 4.2 : *Relative frequency distributions of exterior*

Source and tributary source link-lengths at Perth Amboy together with the fitted log-normal (solid line) and gamma (dashed line) distributions. The class interval is equal to one quarter of the medium exterior link-length. The number besides each distribution indicate the sample size and mean link-length in metres.

Relative frequency distributions for the exterior, *S* and *TS* link-lengths are reproduced in Fig. 4.2.

Eastern Kentucky

Krumbein and Shreve (1970) delineated valley networks over an area of approximately 650 km^2 in Eastern Kentucky on the basis of contour crenulations on U.S. Geological Survey topographic maps with a scale of 1:24,000 and a contour interval of 40 ft. Exterior link-lengths were measured directly on the maps as the straightline distance between the ends of each link. These data have been tabulated by the authors (1970, p. 112, 116-117) for 153 magnitude 5 networks located in the Inez quadrangle, and for 30 magnitude 10 networks selected from the entire study area. Regrettably, it is not possible to distinguish between *S* and *TS* link-lengths in the original publication, so only the two samples of exterior link-lengths are analyzed in this study.

Blountville, Pennington and Copper Ridge

These areas, which are located in northeastern Tennessee and southwestern Virginia, were investigated by Miller (1953, p. 6-7, 16-19, 21-22) as examples of maturely dissected landscapes with dendritic drainage systems developed in homogeneous lithologies. The homoclinal ridge of Clinch Mountain, which was the main area examined by Miller, has not been included in this study because the exterior link-lengths on one flank of the ridge are clearly influenced by the geological structure (Miller 1953, p. 14). The channel networks in the Blountville, Pennington and Copper Ridge areas were mapped from Tennessee Valley Authority topographic maps at a scale of 1:24,000; the accuracy of the mapped networks was checked both by field work and by the study of air photographs. Unfortunately, the lengths of the exterior links in these three areas are tabulated by Miller [1953, Table 2 (1)] in class and these tables are particularly limited where the parameters of the distribution have been estimated from the data, as in the present study. What is more, little or no work appears to have been done on the application of these tests to grouped data, and some of the data used in this study wre eonly available in this form.

The results of the χ^2 goodness-of-fit tests are summarized in Table 4.2, and the fitting of the log-normal and gamma densities to a number of sets of data is illustrated in Figs. 4.1 and 4.2. The log-normal density satisfactorily fits 7 of the 12 exterior link-length distributions, 3 of the 7 *S* link-length distributions, and 5 of the 7 *TS* link-length distributions. On the other hand, the gamma density fits only 3 of the 12 exterior link-length distributions, 2 of the 7 *S* link-length distributions and 2 of the 7 *TS* link-length distributions. Thus, of the 26 link-length distributions examined, 15 are adequately fitted by the log-normal density and 7 by the gamma. In addition, the calculated χ^2 value for the log-normal fit is less than that for the gamma for all but three of the 26 link-length distributions.

Two conclusions may be drawn from these results: (1) *S* and *TS* link lengths do not conform to radically different probability density functions; and (2) the log-normal density generally affords a better empirical fit than the gamma density to exterior, *S* and *TS* link-lengths alike. Still, 11 of the 26 observed link-length distributions depart significantly from the fitted log-normal distributions. Moreover, a high proportion of the fitted log-normal distributions, including many that are statistically acceptable, tend to deviate in frequencies, and there is no way of distinguishing between *S* and *TS* link-lengths.

Analysis

Goodness-of-fit tests

In the absence of a satisfactory theoretical model, investigators have hitherto fitted either the log-normal or gamma density to exterior link-length data collected from natural channel or valley networks. The main reason for choosing these densities appears to have been that they were approximately the correct shape. Yet a number of authors, including Strahler (1954), Schumm (1956) and Krumbein and Shreve (1970), have failed to report goodness-of-fit tests, and these densities have never been compared to see which provides the better fit to a set of data. Therefore, as a first step in this study, both the log-normal and gamma densities were fitted by the maximum

TABLE 4.2 : **Log-normal and gamma distributions fitted to exterior, source and tributary-source link-lengths**

Area	Link Type*	Degrees of Freedom	Log-normal Distribution		Gamma Distribution	
			χ^2	Decision** ($\alpha=0.05$)	χ^2	Decision*** ($\alpha=0.05$)
1	2	3	4	5	6	7
Fowlers Gap	E	16	22.07	Accepted ($0.20>P>0.10$)	122.29	Rejected ($0.001>>P$)
	S	13	22.57	Rejected ($0.05>P>0.025$)	97.74	Rejected ($0.001>>P$)
	TS	16	25.78	Accepted ($0.10>P>0.05$)	44.86	Rejected ($0.001>P$)
Wollondilly	E	10	18.04	Accepted ($0.10>P>0.05$)	20.54	Rejected ($0.025>P>0.02$)
	S	9	9.35	Accepted ($0.50>P>0.30$)	12.02	Accepted ($0.25>P>0.20$)
	TS	9	10.17	Accepted ($0.50>P>0.30$)	8.09	Accepted ($0.70>P>0.50$)
Connors Range	E	12	33.91	Rejected ($0.001>P$)	180.82	Rejected ($0.001>>P$)
	S	9	14.96	Accepted ($0.10>P>0.05$)	54.47	Rejected ($0.001>P$)
	TS	9	27.89	Rejected ($0.005>P>0.001$)	73.83	Rejected ($0.001>P$)

(Contd.)

TABLE 4.2—*contd.*

1	2	3	4	5	6	7
Koetong	E	13	16.57	Accepted (0.25>P>0.20)	35.25	Rejected (0.001>P)
	S	10	20.98	Rejected (0.025>P>0.02)	33.51	Rejected (0.001>P)
	TS	12	11.64	Accepted (0.50>P>0.30)	30.34	Rejected (0.005>P>0.001)
Bobadah	E	14	31.68	Rejected (0.005>P>0.001)	142.33	Rejected (0.001>>P)
	S	9	23.07	Rejected (0.01>P>0.005)	55.19	Rejected (0.001>P)
	TS	14	19.36	Accepted (0.20>P>0.10)	51.49	Rejected (0.001>P)
Goulds Country	E		21.09	Rejected (0.025>P>0.02)	101.87	Rejected (0.001>>P)
	S	7	22.17	Rejected (0.005>P>0.001)	59.16	Rejected (0.001>P)
	TS	10	11.91	Accepted (0.30>P>0.25)	42.44	Rejected (0.001>P)
Perth Amboy	E	8	6.53	Accepted (0.70>P>0.50)	3.71	Accepted (0.90>P>0.80)
	S	5	7.80	Accepted (0.20>P>0.10)	9.90	Accepted (0.10>P>0.05)
	TS	7	15.24	Rejected (0.05>P>0.025)	8.09	Aceepted (0.50>P>0.30)

Eastern Kentucky (magnitude 5 basins)	E	20	42.13	Rejected (0.005>P>0.001)	159.68	Rejected (0.001>>P)
Eastern Kentucky (magnitude 10 basins)	E	6	24.36	Rejected (0.001>P)	30.01	Rejected (0.001>P)
Blountville	E	3	7.42	Accepted (0.10>P>0.05)	13.50	Rejected (0.005>P>0.001)
Pennington	E	1	0.98	Accepted (0.50>P>0.30)	1.02	Accepted (0.50>P>0.30)
Copper Ridge	E	1	0.48	Accepted (0.50>P>0.30)	1.04	Accepted (0.50>P>0.30)

*E=exterior links, S=source links, TS=tributary-source links.

**H_o : The sample of exterior (S or TS) link-lengths was randomly selected from a log-normally distributed population of such lengths.

*** H_o : The sample of exterior (S or TS) link-lengths was randomly selected from a gamma distributed population of such lengths.

likelihood method (Aitchison and Brown 1966, p. 39; Greenwood and Durand 1960) to the exterior, *S* and *TS* link-length data from the 11 study areas. (Only the gamma density with a positive shape parameter was considered as a plausible link-length model since a negative shape parameter implies that the relative frequency of links approaches infinity as link-length approaches zero.) The adequacy of the fitted distributions was evaluated by the usual χ^2 goodness-of-fit test. Although this test is generally not very powerful for continuous distributions (Green and Hegazy 1976; Stephens 1974), other tests, such as the Cramer-von Mises, Anderson-Darling and Shapiro—Wilk tests, have only been tabulated for limited cases such as a testing for the goodness-of-fit of normal, exponential and uniform distributions; the same way from the observed link-length distributions. This tendency is evident in the fact that the modal frequency is underestimated by the fitted log-normal distributions in 20 of the 26 observed distributions (see for example Fig. 4.1). Thus, although the log-normal density is superior to the gamma, it can clearly be improved upon in the search for a satisfactory model for exterior, *S* and *TS* link-lengths.

Graphical analysis

A second method was, therefore, employed to learn more about the distributions of these link-lengths. This method (Hahn and Shapiro 1967, pp. 195-224; Johnson and Kotz 1970, pp. 9-30) makes use of the fact that many probability densities, including the log-normal and gamma densities, can be adequately described by four statistics, namely, the mean μ, the standard deviation σ, the standardized measure of skewness β_1, and the standardized measure of kurtosis β_2 (Bliss 1967, pp. 142-143). For the log-normal and gamma densities the relationship betweeu σ/μ (the coefficient of variation), β_1 and β_2 can be specified algebraically (Hahn and Shapiro 1967, pp. 124, 128), and each relationship represented graphically by a curve on a diagram in which one statistic is graphed against another. The three resulting diagrams may then be used to obtain in indication of whether or not a given set of link-length data can be adequately represented by either a log-normal or gama distribution. This is done by calculating the sample estimates of

σ/μ, β_1 and β_2 which are denoted by s/m, b_1 and b_2 respectively, and by plotting these sample statistics on the indicated diagrams. If the sample of link-lengths were drawn from either a log-normal or a gamma distributed population, the point plotted on each diagram should fall close to the curve representing the relevant probability density.

The estimates of b_1 and b_2 were calculated using $b_1=g_1^2$ and $b_2=g_2+3$ (Fisher 1958, p. 75; Bliss 1967, p. 144). The bias corrections embodied in g_1 and g_2 remove virtually all of the bias in estimating β_2, but a small positive bias remains in the estimate of β_1. The expression for this residual bias is quite complex, involving high-order moments of the distribution, and for this reason was not used here. For practical purposes, however, this bias is negligible, and does not materially affect the present results.

There are two limitations which should be borne in mind when applying the graphical method outlined above. Firstly, the sample statistics s/m, b_1 and b_2 are only estimates of σ/μ, β_1 and β_2 and are subject to sampling fluctuations; b_1 and b_2, in particular, are very sensitive to a few extreme observations. Secondly, the shape of a distribution is not in general uniquely defined by any pair of the statistics σ/μ, β_1 and β_2. Therefore, because a sample of link-lengths plots close to the curve for a particular density on one of the diagrams, it does not necessarily mean that the data will be adequately fitted by that density (Hahn and Shapiro 1967, p. 198).

Diagrams for β_1 *versus* σ/μ and β_2 *versus* σ/μ on which are graphed the relationship between the relevant pair of statistics for the log-normal and gamma densities are depicted in Fig. 4.3 and 4.4. The sample statistics for the 12 samples of exterior link-lengths and for the 14 sub-samples of *S* and *TS* link-lengths (Table 4.3) are also plotted on these diagrams. The plotted points tend to scatter around the log-normal rather than the gamma curve, thereby, confirming the above conclusion that the log-normal is generally superior to the gamma density as an empirical model for exterior *S* and *TS* link-lengths alike. Figs. 4.3 and 4.4 show that the main failing of the gamma model is that a gamma distribution with the same coefficient of variation as a link-length distribution tends to underestimate both the skewness and kurtosis of the link-length

TABLE 4.3 : **Statistics for exterior, source and tributary-source link-length distributions**

Area	Link Type*	Sample Size	Mean (metres)	Standard Deviation (metres)	Standardized Measure of Skewness	Standardized Measure of Kurtosis
Fowlers Gap	E	1031	0.791	0.681	11.47	23.36
	S	524	0.717	0.638	10.60	20.01
	TS	507	0.867	0.715	12.28	25.53
Wollondilly	E	1040	332.0	175.6	1.13	4.23
	S	490	308.0	170.9	1.45	4.81
	TS	550	353.5	177.1	0.95	3.92
Connors Range	E	1245	78.5	46.6	4.38	10.45
	S	600	73.5	46.8	6.29	14.49
	TS	645	83.1	46.0	3.13	7.20
Koetong	E	753	312.3	193.2	2.24	5.74
	S	329	277.3	182.4	3.21	7.80
	TS	424	339.5	197.0	1.83	4.82
Bobadah	E	685	194.8	161.0	8.38	14.80
	S	372	146.8	120.2	17.60	30.92
	TS	313	251.7	183.5	5.55	10.34
Goulds Country	E	1100	91.8	50.9	2.93	7.08
	S	534	78.3	41.2	2.94	7.06
	TS	566	104.5	55.7	2.50	6.33

Perth Amboy	E	214	3.20	1.91	0.84	3.57
	S	96	2.52	1.71	2.69	5.88
	TS	118	3.76	1.88	0.37	3.35
Eastern Kentucky (magnitude 5 basins)	E	765	139.6	64.6	4.50	10.58
Eastern Kentucky (magnitude 10 basins)	E	300	169.1	75.6	3.56	9.00
Blountville	E	198	170.3	88.7	2.46	6.94
Pennington	E	50	160.9	71.3	0.41	3.25
Copper Ridge	E	50	223.7	97.6	2.27	6.55

* E = exterior link, S = source link, TS=tributary-source link.

distribution. This failing is particularly marked where link-length distributions have high coefficients of variation, and is evident in Figure 4.1 where the fitted log-normal distributions fill out both the peak and tail of the observed link-length distributions far better than do the fitted gamma distributions.

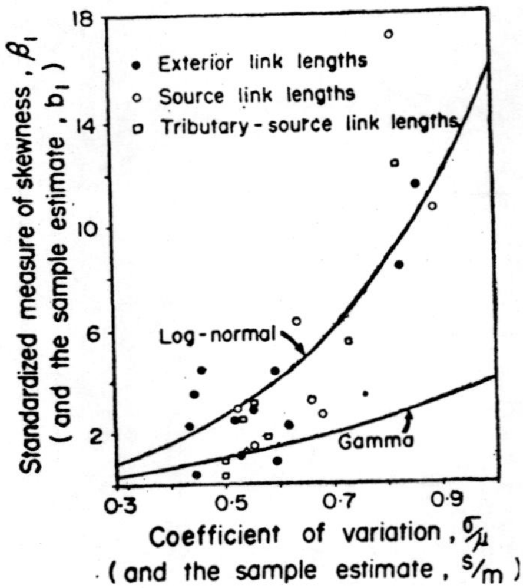

FIG. 4.3 : *Plot of the standardized measure of skewness against the coefficient of variation for the 26 exterior.*

Source and tributory-source link-length distributions under study.

The diagram for β_1 *versus* β_2 appears in Fig. 4.5. On this diagram are plotted b_1 and b_2 for the 26 link-length distributions under study (Table 4.3), along with the relationships between β_1 and β_2 for the log-normal and gamma densities. Three findings emerge from this diagram :

(1) As was evident in Figs. 4.3 and 4.4, the gamma density (with a positive shape parameter) has too limited a range of skewness and kurtosis to adequately fit many of the observed link-length distributions.

(2) Although the log-normal density provides a satisfactory fit to most of the observed exterior, *S* and *TS* link-length distributions over virtually the full range of observed b_1 and b_2

(Table 4.2), these distributions always plot above (or to the left) of the log-normal curve and, hence, appear to depart from the log-normal model in a systematic and consistent manner.

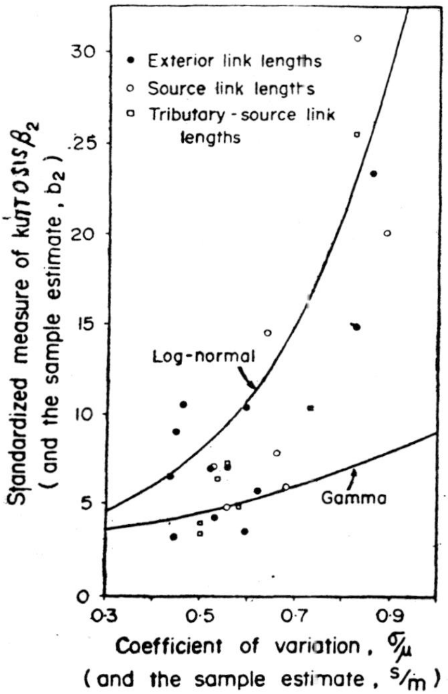

FIG. 4.4 : *Plot of the standardized measure of kurtosis against the coefficient of variation for the 26 exterior.*

Source and tributary-source link-length distributions under study.

(3) Regardless of whether the points on the diagram represent exterior, *S* or *TS* link-lengths, they exhibit a linear (or perhaps near-linear) trend with remarkably little scatter. This trend suggests that the observed exterior, *S* and *TS* link-length distributions all approximate to varying degrees a single underlying probability density.

Discussion

Mapping and measurement errors

The exterior, *S* and *TS* link-length distributions represented by the points plotted in Figs. 4.3 4.4 and 4.5 relate to measured link-lengths which may differ from actual link-length owing to (1) errors in mapping channel network, and (2) errors in measuring link-lengths. Mapping errors are likely to occur where channel networks are delineated on the basis of some criteria other than the field definition of a channel, whereas measurement errors are commonly introduced in a variety of ways, notably in the present context by link-lengths being measured along straight lines or in the horizontal plane. Because mapping and/or measurement errors are probably present to some extent in all of the data analyzed in this study, the possibility must be considered that the consistent manner in which the observed link-length distributions depart from the log-normal density can be attributed in large part to these errors.

In all but two of the study areas, channel or valley networks were mapped and link-lengths measured on either topographic maps or air photographs. Although most of the mapped networks have been checked to varying degrees in the field, there are probably significant mapping and measurement errors in the link-length data from these areas. Because these errors are complex and their severity is difficult to estimate, it is virtually impossible to quantitatively evaluate their impact on the observed exterior, *S* and *TS* link-length distributions. At Perth Amboy, the channel network was mapped by field survey; but since link-lengths were measured on the map without correcting for gradient, measurement errors have been introduced into the link-length data. At Fowlers Gap, however, channel networks were mapped in the field and link-lengths were measured along the ground. Thus, any errors in the link-length data from this area are probably negligible, and the distributions of measured exterior, *S* and *TS* link-lengths can be expected to closely represent the distributions of actual lengths.

An examination of Fig. 4.5 reveals that the Fowlers Gap distributions all plot well above the log-normal curve, and more or less along the same line (or curve) as the distributions for the other study areas. The first of these observations confirms that actual exterior, *S* and *TS* link-lengths are not log-normally distributed, while the second suggests that the trend evident in the points plotted in Fig. 4.5 has been little affected by the various errors embodied in the observed link-length distributions. These errors, however, have presumably added to the scatter of points around the trend.

The latter conclusion is in keeping with the results of a recent study by Smart (1978) in which 30 channel networks in eastern Kentucky were outlined on topographic maps using the usual contour-crenulation criterion in one instance and a channel-slope criterion in the other. Smart (1978, p. 167) found a remarkable similarity between the properties of the two types of networks, and concluded that it is possible that the factors which control channel network properties act at all scales, so that networks mapped according to different criteria have (amongst other things) the same general kinds of link-length distributions. Both the present study and that of Smart highlight the need for further work on the effect of errors introduced by collecting link-length data from maps. This need deserves immediate attention for the large sample sizes required for the meaningful statistical analysis of such data make it increasingly impracticable to map and measure link-lengths in the field.

A single underlying probability density

Perhaps the most surprising of the present findings is that, regardless of the differences in climate, geology, relief and stage of landscape development amongst the study areas, the observed exterior *S* and *TS* link-length distributions all appear to approximate to varying degrees a single underlying probability density. This conclusion is implied by the strong linearity (or curvilinearity) in the cluster of points in Fig. 4.5, and by the weaker linearity (or curvilinearity) in Figs. 4.3 and 4.4. Although an attempt was made using the observed relationships between b_1 and b_2 and s/m (Figs. 4.3, 4.4 and 4.5) to derive a moment-generating function corresponding to a simple analytical expression for the underlying probability density, it was without success, and the precise form of the underlying density remains unknown. Attempts were also made to identify the geomorphic mechanism giving rise to this density. Several hypotheses were tested in the course of which it became clear that the scatter of points in Figs. 4.3, 4.4 and 4.5 is not due entirely to sampling variation and mapping and measurement errors, but also to variations in relief (see Abrahams 1977). However, the geomorphic mechanism giving rise to the underlying probability density remains an enigma. Work is continuing on this project, and recently completed analyses suggest that interior link-lengths conform to the same underlying probability density. This finding reinforces the notion that the mechanism giving rise to this density is of a very general nature; but the identity of this mechanism has still to be determined.

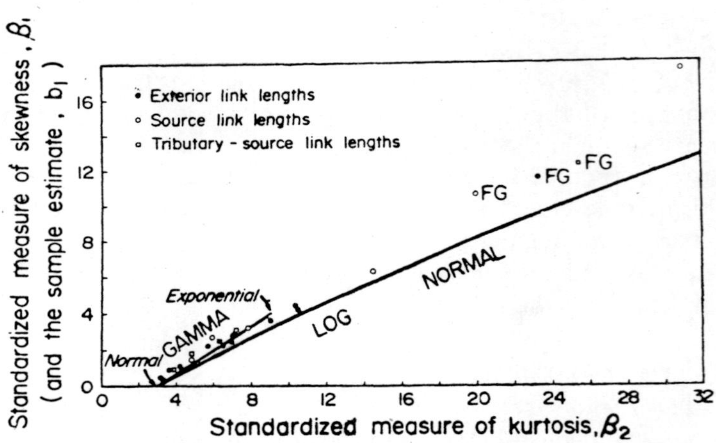

FIG. 4.5 : *Plot of the standardized measure of skewness against the standardized measure of kurtosis for the 26 exterior.*

Source and tributary : source link-length distributions under study. The distributions for Fowlers Gap are denoted by *FG*.

CONCLUSION

Although the graphical method of examining frequency distributions has severe limitations, it allows a number of valuable insights into the nature of the distributions of exterior, *S* and *TS* link-lengths. The graph of β_1 against β_2 is particularly useful and suggests that these link-lengths conform to varying degrees to a single underlying probability density. Unfortunately, it has not been possible to determine either the precise form of the underlying density or the geomorphic mechanism giving rise to it. The log-normal density, however, affords a reasonably close approximation to the density and, hence, to most sets of exterior, *S* and *TS* link-length data. Although the gamma density also provides a satisfactory fit to some link-length distributions, it is totally inadequate where such distributions have high measures of skewness and kurtosis. Moreover, even where the gamma fit is satisfactory, the log-normal fit is usually better. Thus, the log-normal density is clearly superior to the gamma as an empirical model for exterior, *S* and *TS* link-lengths; and until more is known of the underlying probability density for these lengths, it should be used in preference to the gamma in this role.

SUGGESTED READINGS

Abrahams, A.D., 1972 : Factor analysis of drainage basin properties : Evidence for stream abstraction accompanying the degradation of relief, *Water Res. Research*, V.8, p. 624-633.

Abrahams, A.D., 1977 : The factor of relief in the evolution of channel networks in mature drainage basins, *Amer. Jour. Sci.*, V. 277, pp. 626-646.

Abrahams, A.D. and Campbell, R.N., 1976 : Source and tributary-source link-lengths in natural channel networks, *Bull. Geol. Soc. Amer.*, V. 87, pp. 1016-1020.

Aitchison, J. and Brown, J.A.C., 1966 : *The log-normal distribution*, Cambridge Univ. Press, Cambridge, p. 176.

Bliss, C.I., 1967 : *Statistics in biology*, McGraw-Hill Book Co., New York, pp. 558.

Dunkerley, D.L., 1977 : Frequency distributions of stream link lengths and the development of channel networks, *Jour. Geology*, V. 85, pp. 459-470.

Fisher, R.A., 1958 : *Statistical Methods for Research Workers* (13th ed.). Oliver and Boyd, Ediburgh, p. 356.

Green, J.R. and Hegazy, Y.A.S., 1976 : Powerful modified-EDF goodness-of-fit tests, *Jour, Amer. Statist. Assoc.*, V. 71, pp. 204-209.

Greenwood, J.A. and Durand, D., 1960 : Aids for fitting the gamma distribution by maximum likelihood, *Technometrics*, V. 2, pp. 55-65.

Hahn, G.J. and Shapiro, S.S., 1967 : *Statistical models in engineering*, John Wiley and Sons, New York, p. 355.

Johnson, N.L. and Kotz, S., 1970 : *Continuous univariate distributions*, 1, Houghton Mifflin Co., Boston, p. 300.

Krumbein, W.C. and Shreve, R.L., 1970 : Some statistical properties of dendritic channel networks, *Off. Nav. Research Task 389-150, Tech. Rept.* 13; *Nat. Sci. Found. Grant GA-1137 Spec. Proj. Rept.*, p. 117.

Leopold, L.B. and Langbein, W.B., 1962 : The concept of entropy in landscape evolution, *U.S. Geol. Survey Prof. Paper 500-A*, p. A1-A20.

Mark, D.M., 1975 : Geomorphometric parameters : A review and evaluation, *Geogr. Annaler*, 57A, p. 165-177.

Maxwell, J.C., 1960 : Quantitative geomorphology of the San Dimas experimental forest, California. *Off. Nav. Research Proj. 389-042, Tech. Rept.* p. 19, 95.

Miller, T.C., 1953 : A quantitative geomorphic study of drainage basin characteristics in the Clinch Mountain area, Virginia and Tennesee, *Off. Nav. Research Proj. 389-042, Tech. Rept. 3, p. 30*

Mock, S.J., 1971 : A classification of channel links in stream networks, *Water Res. Research*, V. 7, p. 1558-1566.

Schumm, S.A., 1956 : Evolution of drainage systems and slopes in badlands at Perth Amboy, New Jersey. *Bull. Geol. Soc. Amer.*, V. 67, pp. 597-646.

Shreve, R.L., 1966 : Statistical law of stream numbers, *Jour. Geology*, V. 74, p. 17-37.

Shreve, R.L., 1967 : Infinite topologically random channel networks, *Jour. Geology*, V. 75, pp. 178-186.

Shreve, R.L., 1969 : Stream lengths and basin areas in topologically random channel networks, *Jour. Geology*, V. 77, pp. 397-414.

Smart, J.S., 1972 : Channel networks, *Advances in Hydroscience*, V. 8, p. 305-346.

Smart, J.S., 1978 : The analysis of drainage network composition, *Earth Surface Processes*, V.3, pp. 129-170.

Surkan, A.J. and Considine, J.P., 1967 : Digital simulation of channel networks, *Internat. Assoc. Sci. Hydrology Pub.* 75, p. 87-98.

Stephens, M.A., 1974 : EDF statistics for goodness-of-fit and some comparisons, *Jour. Amer. Statist. Assoc.*, V. 69, p. 730-737.

Strahler, A.N., 1954 : Statistical analysis in geomorphic research. *Jour. Geology*, V. 62, p. 1-25.

FIVE

John F. Orsborn, Nadira Kabir and *F. David Deane*

RELATIONSHIPS AMONG LOW, AVERAGE AND FLOOD FLOWS

INTRODUCTION

Rationale

ALTHOUGH hydrologists have developed numerous models and methods for estimating low, average and flood flows at ungaged sites on streams 1, 2, 4, 5, there are several limitations on all methods :

(1) Some require extensive and expensive parametric data collection systems.
(2) Some require the determination of variables which at best can be considered only as approximations.
(3) Only a few of the methods have more than a very limited geographic or hydrologic range of application.
(4) Some of the parameters do not have stable physical relationships with the unknown flow.
(5) Even fewer of the methods can be used to relate one type of streamflow (low, average or flood) to another.

Rarely is a stream gage located near a point of interest on the stream at the time of need. The results of this project will help provide solutions to this type of problem based on factual data and physical relationships. The model described here relates one characteristic flow to another (output : output) and considers the entire drainage basin as the integrator of its streamflows. Provincial correlations for average and low flows for gaged streams and various combinations of geomorphic parameters of the drainage basins above the gages were developed.

To determine these flows from the ungaged watersheds, the same geomorphic parameters are determined and entered into the provincial correlations, thus, yielding the desired ungaged flows. Using the provincial correlations as developed here between floods, average and low flows, flood flows in ungaged streams can then be predicted.

The results of this project will be useful in defining water availability. By using the output-output method and other hydrogeomorphic methods for determining low and average flows in ungaged basins, it is possible to get a good idea of the natural flow regime of any basin, and man-made influences on that regime, with a minimum of information about the basin. The model can be used to : (1) increase the predictive capability of hydrologic analyses of ungaged streamflows; (2) determine frequency and duration curves at ungaged stream sites; and (3) provide more economical designs of culverts, bridges and other water conveyance structures involving expenditures of public funds. If one considers that uncertainty in hydrologic data increases with the distance upstream from a gaging station, and also as a function of the number of side streams passed and their cumulative flows, usually one would not have to move very far before confidence in transposing gaging records would be markedly decreased.

Research procedures

Long-term records from gaging stations which are minimally affected by regulation or diversion were selected for the analysis from each hydrologic province in Washington, Idaho and Oregon. The hydrologic province divisions selected for this study on the basis of climatic uniformity are shown in Fig. 5.1.

A computerized procedure was utilized to process the data perform mathematical regressions for each hydrologic province and arrange the flood prediction equation in the desired form. A comparative analysis of the provincial equations was made to explore how equation coefficients varied from province to province. Some provincial regression analyses were strongly influenced by having too few data with which to perform the regression. A semi-empirical, graphical technique for deriving the provincial equations was developed using only small samples of the data. This method performs well with fewer data points than are required for the mathematical regressions.

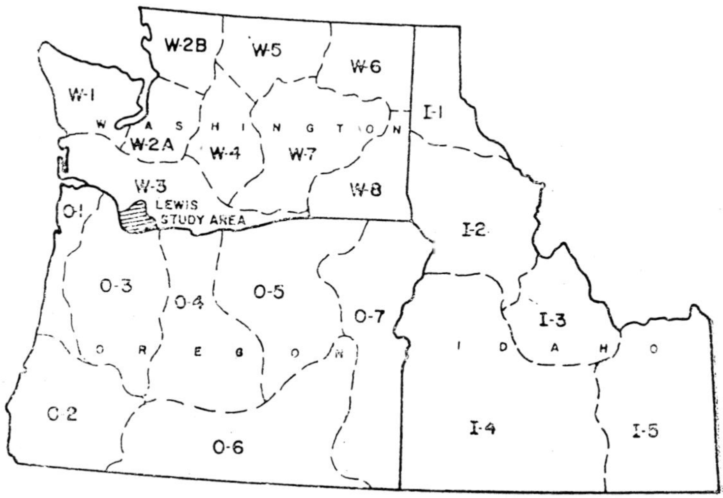

FIG. 5.1 : *Hydrologic provinces for low, average and flood flow analysis in Washington, Oregon and Idaho.*

Definition of streamflow symbols

The following symbols are used throughout this report to designate the various streamflows considered. All flow terms have units of cubic metres per second (cms). Log-Pearson, Type III recurrence interval values were used for floods and low flows.

QAA : Average annual flow; the mean of all annual mean flows for the period of record.

Q7L2 : 7-day, 2-yr. low flow; 7-day average low flow with a 2-yr. recurrence interval.

QF2P : Instantaneous peak flood with a recurrence interval of 2yrs. based on annual peak flood series.

QF50P : Instantaneous peak flood with recurrence interval of 50 yrs. based on annual peak flood series.

QF2D : Average daily flood with a recurrence interval of 2 yrs. based on annual maximum day flood series.

QF50D : Average daily flood with a recurrence interval of 50 yrs. based on annual maximum day flood series.

Correlations between characteristic flows and watershed parameters

The low and average flows of gaged stations of perennial streams are related to basin parameters. In the Lake Coeur d'Alene watershed in northern Idaho, for example, the following prediction formula for Q7L2 is applicable :

$$Q7L2 = 23.1\ (LT\ .\ H^{0 \cdot 5}) \qquad \ldots\ldots(1)$$

In which LT = total length of streams in km; and H = basin relief in km.

Regional or provincial relationships between average annual streamflow and the average annual volume of precipitation on a watershed tend to be highly correlated. In the equation form :

$$QAA = C_1(P.A) \qquad \ldots\ldots(2)$$

in which C_1 = a runoff coefficient which varies as a function of climatic region and is larger for areas with greater values P; P= the average annual precipitation; and A = the watershed drainage area. Actual values of C_1vary between 2.8×10^{-4} in humid areas with more than 200 cm of precipitation per year to about 0.65×10^{-4} for desert areas with 12 to 25 cm of precipitation per year. C_1 has a maximum value of 3.2×10^{-4} (0.0737 in *EGS* units) for an upper limit of average annual runoff caused by 1.0 mm of precipitation per year falling on a watershed of 1.0 square kilometre.

PROJECT BACKGROUND AND ANALYSIS

Regression analysis

The Lewis Study Area in Hydrologic Province 3 in south-western Washington (Fig. 5.1) was selected for model development. The basins are hydrologically and geologically diverse, with some basins experiencing low, low flows and high floods, and others having higher than average low flows. The necessary data and derived numbers for the Lewis Study Area appear in Table 5.1. These new correlations for predicting flows make use of the idea that flood flows, low flows and average flows can be expressed as functions of one another. It was observed during one test that three streams with the same average annual flow had widely varying low flows as noted by points, 1, 2, 3, in Fig. 5.2. Fig. 5.3 is a plot of flood flows $QF2P$ and $QF50P$ with average annual flows at the gaging stations in Table 5.1. The vertical ordering of points 1, 2, and 3 is reversed from that of Fig. 5.2, showing high, low flows correspond to low flood flows and low, low flows correspond to high flood flows. From the relative plotting positions in Figs. 5.2 and 5.3, it is evident that flood flows cannot be expressed as a linear function of low flow. After considering that the several basins with the same average annual runoff (but with different low flows) must have differences in geology and, therefore, surface runoff, Figs. 5.4a and 5.4b were developed showing a relationship of floods and low flows to average flows. By multiplying each average low flow at each gage by its 2-yr. and 50-yr. recorded peak flood flows squared, the very strong correlations in Figs. 5.5a and 5.5b appeared. The scattered points 1, 2 and 3 in Figs. 5.2 are clustered at the lower end of the graphs in Figs. 5.4a and 5.4b. In the equation form, Figs. 5.4a and 5.4b are given by Eqs. 3 through 6 :

$$(QF2P)^2\,(Q7L2) = 8.31\,(QAA)^{3\cdot 08} \qquad \ldots\ldots(3)$$

$$QF2P = 2.88\left[\frac{(QAA)}{(Q7L2)^{0\cdot 32}}\right]^{1\cdot 54} \qquad \ldots\ldots(4)$$

and

$$(QF50P)^2\,(Q7L2) = 61.98\,(QAA)^{2\cdot 93} \qquad \ldots\ldots(5)$$

or

$$QF50P = 7.87\left[\frac{(QAA)}{(Q7L2)^{\cdot 34}}\right]^{1\cdot 47} \qquad \ldots\ldots(6)$$

To test the relationships in Figs. 5.4a and 5.4b for several regions in Washington, similar plots were made. The graphical interpretations of the regression lines consistently produced results in the form of Eq. 3 as :

$$(QF2)^2\,(Q7L2) = a\,(QAA)^{3\cdot 00} \qquad \ldots\ldots(7)$$

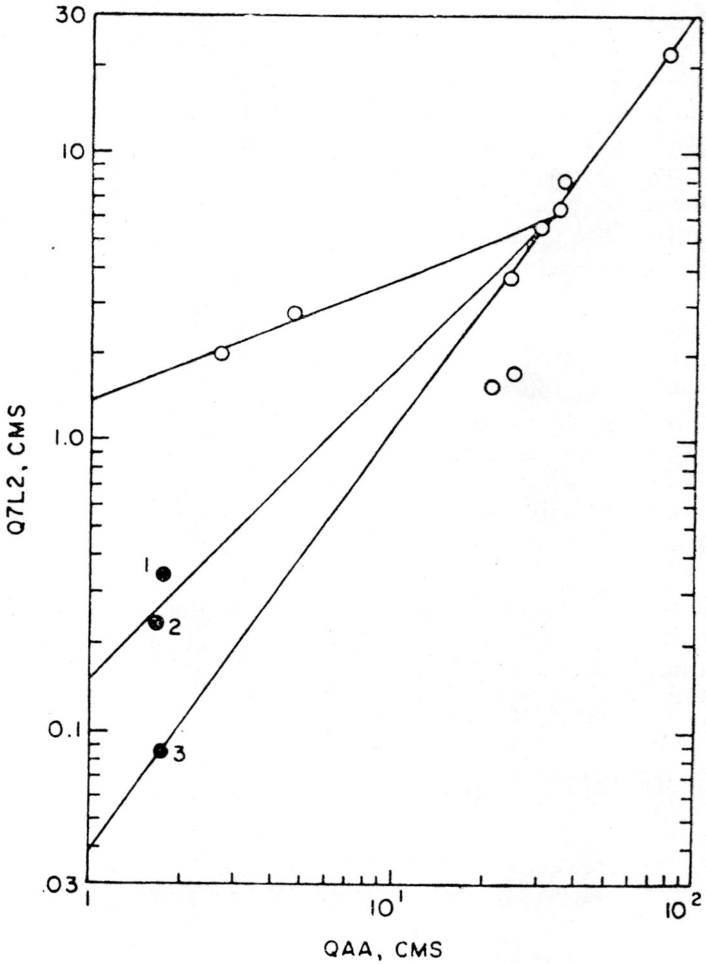

FIG. 5.2 : *Low flow related to average flow*

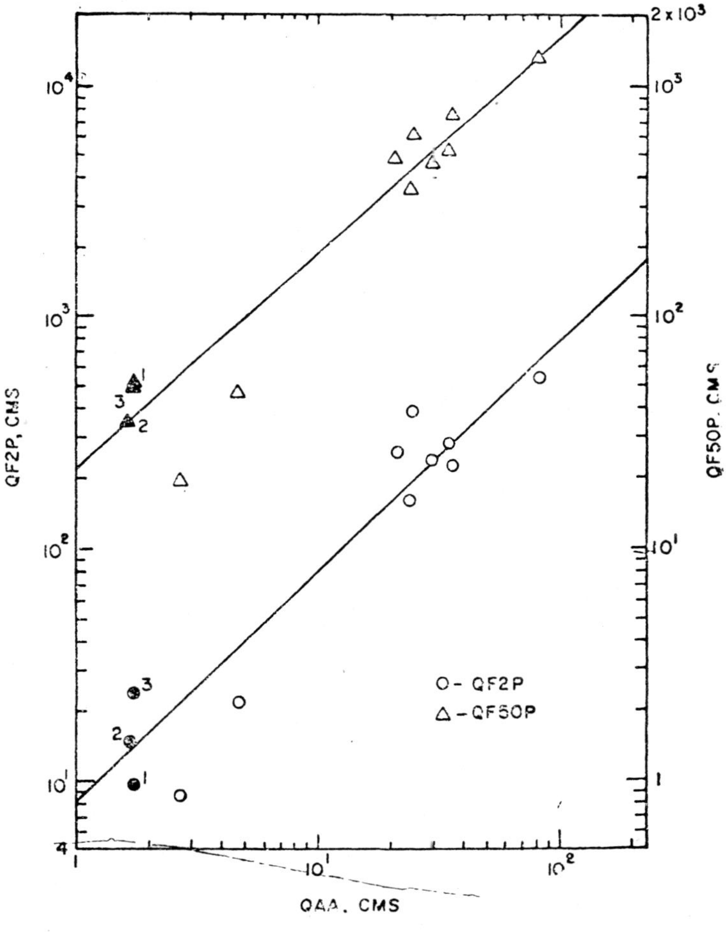

FIG. 5.3 : *Flood flow related to average flow*

TABLE 5.1 : **Lewis river basin sample stations data and derived numbers for analytical development**

Station	Number	QF50P (cms)	QF2P (cms)	QAA (cms)	Q7L2 (cms)	QF50P / Q7L2	QF2P / Q7L2	QAA / Q7L2	$(QF50P)^2$ × 10( )	$(QF2P)^2$ × 10( )	Period of Record
Washougal R.	14-1435	614.1	384.9	24.68	1.73	356.0	223.0	14.30	$3.77(10)^5$	$1.48(10)^5$	1944-68
Salmon Cr. (3)	-2120	48.7	24.0	1.73	0.09	573.0	282.0	20.30	$2.37(10)^3$	$5.74(10)^2$	1943-68
Big Cr. (2)	-2135	35.7	14.7	1.67	0.23	154.0	63.3	7.20	$1.27(10)^3$	$2.14(10)^2$	1927-31 1955-68
Meadow Cr.	-2145	19.2	8.6	2.69	2.01	9.6	4.3	1.34	$3.70(10)^2$	$7.45(10)^1$	1927-31 1955-65
Rush Cr.	-2150	44.7	21.8	4.70	2.80	16.0	7.8	1.68	$2.00(10)^3$	$4.75(10)^2$	1927-31 1955-62 1963-68
Lewis R.	-2160	735.8	227.8	35.97	7.95	92.5	28.7	4.52	$5.41(10)^5$	$5.19(10)^4$	1927-34 1954-68
Muddy R.	-2165	345.3	157.9	23.97	3.76	91.7	42.0	6.36	$1.19(10)^5$	$2.53(10)^4$	1927-34 1954-68
Lewis R.	-2180	1307.5	529.2	81.73	22.36	58.5	23.7	3.66	$1.71(10)^6$	$2.80(10)^5$	1910-12 1924-58
E. F. Lewis R.	-2225	469.8	254.7	21.11	1.56	302.0	164.0	12.10	$2.21(10)^5$	$6.49(10)^4$	1929-68
Kalama R.	-2230	447.1	236.9	30.00	5.60	80.0	42.3	5.35	$2.00(10)^5$	$5.61(10)^4$	1911-13 1916-32
Kalama R.	-2235	515.1	279.9	34.78	6.45	80.0	43.4	5.40	$2.64(10)^5$	$7.85(10)^4$	1946-71
Curley Cr. (1)	-2155	50.9	9.6	1.75	0.34	150.0	28.4	5.17	$2.59(10)^3$	$9.29(10)^1$	1955-70

Note—Stations (1), (2), and (3) referred to in the text, Fig. 5.2 and elsewhere.

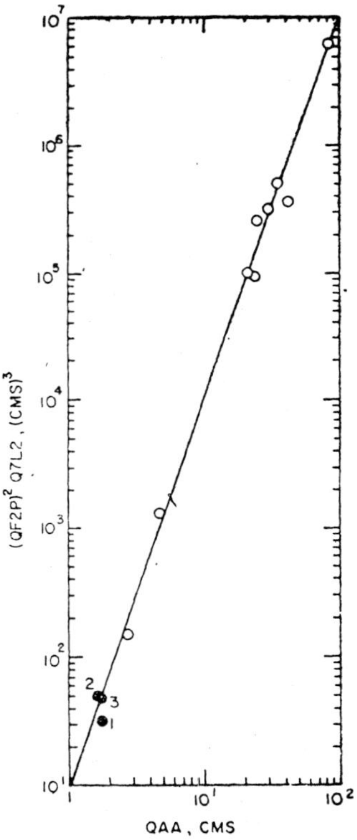

FIG. 5.4*a* : *Average annual flows related to average low flows and 2-yr. floods.*

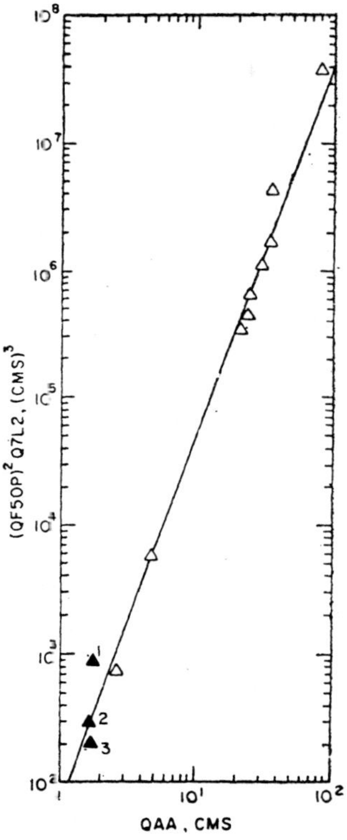

FIG. 5.4*b* : *Average annual flows related to average low flows and 50-yr. floods*

The coefficient (a) was the only term which showed much variation from region to region. The exponent of the righthand term in Eq. 7 did vary between 2.8 and 3.2. The cause for the variation is felt to be due to variability of periods of record within a region. In some cases the exponent of *Q7L2* varies from 0.32 to 0.34, but the variation in the solution resulting from the use of 0.33 is small.

The graphical relationship in Fig. 5.5 allows a more direct determination of the flow relation. The equation for the line in Fig. 5.5 is :

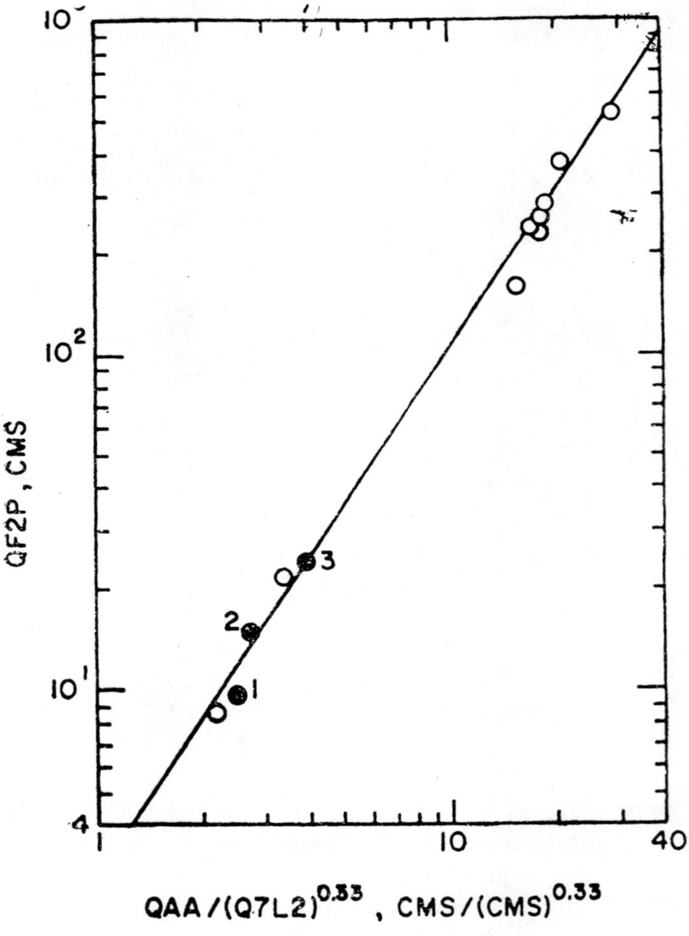

FIG. 5.5 : *Peak flood flow related to a ratio of average flow and low flow.*

$$QF2P=2.79\left[\frac{(QAA)}{(QL72)^{0\cdot 33}}\right]^{1\cdot 56} \qquad \ldots\ldots(8)$$

One would anticipate being able to algebraically divide (QAA) by ($Q7L2$), but as shown in Fig. 5.6 this step does not maintain the original correlation.

Semi-empirical method

As a check on the first method for deriving Eq. 7, a semi-empirical method was developed using a dimensionless duration curve, and by dividing the relationship in Fig. 5.4a into two parts. In this semi-empirical method, component parts of Eq. 7 are developed separately and then combined.

A set of dimensionless average duration curves for the Lewis Study Area is shown in Fig. 5.7. This assumes that $Q7L2$ is exceeded 100 per cent of the time and $QF2$ is exceeded zero per cent of the time, and allows for evaluation of gaging station data points for provincial homogeneity (e.g., two data points lower right in Fig. 5.7). In most cases QAA is exceeded 50 to 60 per cent of the time in Fig. 5.7. This relatively small variation in the per cent of time QAA is exceeded within a region allows use of the ratio $QF2/QAA$ as an indicator of data continuity.

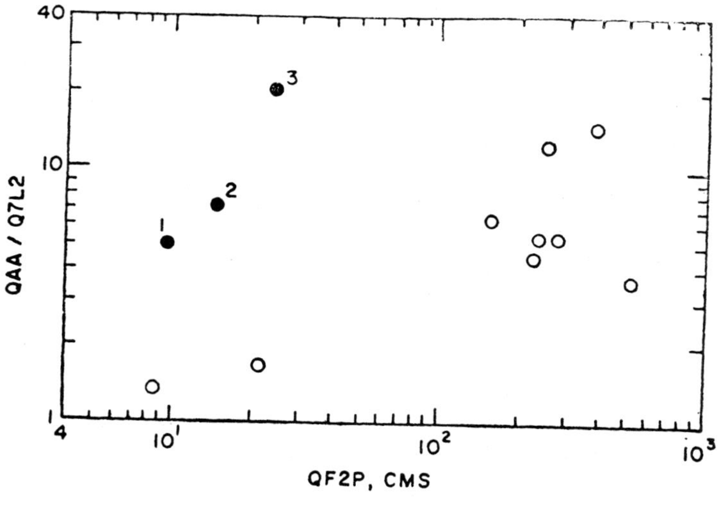

FIG. 5.6 : *Average: low flow ratio related to flood flow*

The ratios of $QF2/Q7L2$ and $QAA/Q7L2$ were related to each other as the first step in the two-part, semi-empirical analysis as shown in Fig. 5.8. A similar relationship also holds for $QF50/Q7L2$. The resulting equation is :

$$\frac{QF2}{Q7L2} = c\left[\frac{QAA}{Q7L2}\right]^n \qquad \text{......(9)}$$

Where (c) and (n) are a regression coefficient and an exponent.

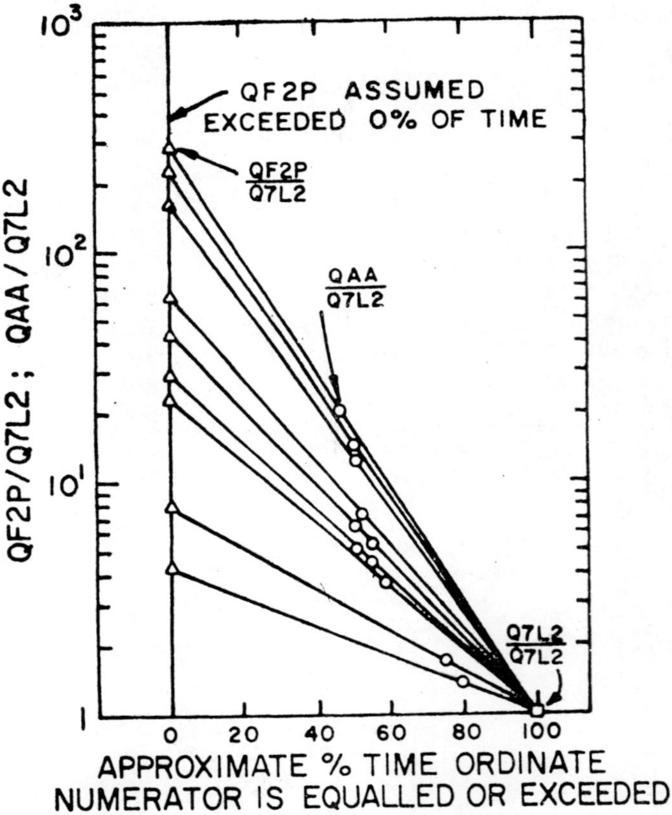

FIG. 5.7 : *Dimensionless duration curve*

Then Eq. 9 can be rearranged in the form:

$$(QF2)^2 (Q7L2)^{2(n-1)}=(c) (QAA)^{2n} \qquad ......(10)$$

It is obvious that if n=1.5, and that if the origin of the theoretical line was at (1,1) for a steady idealized stream, then Eq. 7 is obtained directly. In most cases (n) is about 1.5 and a second plot of $(QF2)^2$ and QAA must be made to complete the analysis. This result is shown for both 2-yr and 50-yr floods in Fig. 5.9 and in the equation :

$$(QF2)^2=c (QAA)^d \qquad ......(11)$$

By substituting the actual values for Eq. 11 from Fig. 5.9 into the actual values for Eq. 9 from Fig. 5.8 and rearranging terms, reducing the exponent of $Q7L2$ to 1.0 and multiplying by Eq. 11, Eq. 14 is derived. Using Eq. 9 with Fig. 5.8 data :

$$QF2P/Q7L2=2.91 (QAA/Q7L2)^{1\cdot 56}$$
$$QF2P=2.91 (QAA)^{1\cdot 56}/(Q7L2)^{0\cdot 56}$$
$$(QF2P)^2=8.47 (QAA)^{3\cdot 12}/(Q7L2)^{1\cdot 12}$$
$$(QF2P)^2 (Q7L2)^{1\cdot 12}=8.47 (QAA)^{3\cdot 12} \qquad ......(12)$$

Using Eq. 11 with Fig. 5.9 data :

$$(QF2P)^2=43.96 (QAA)^{2\cdot 11}$$

Substitute into Eq. 12 and solve for $Q7L2$ in terms of QAA :

$$43.96 (QAA)^{2\cdot 11} (Q7L2)^{1\cdot 12}=8.47 (QAA)^{3\cdot 12}$$
$$(Q7L2)^{1\cdot 12}=0.193 (QAA)^{1\cdot 01}$$
$$Q7L2=0.23 (QAA)^{0\cdot 90}$$

Multiply this result by Eq. 11 from Fig. 9 :

$$(QF2P)^2 (Q7L2)=10.1 (QAA)^{3\cdot 01} \qquad ......(13)$$

Rearrange into form of Eq. 4 :

$$QF2P=3.18\left[\frac{(QAA)}{(Q7L2)^{0\cdot 33}}\right]^{1\cdot 50} \qquad ......(14)$$

The results of a similar analysis using the 50-yr. flood relationships in Fig. 5.8 and 5.9 yield the following equations in the form of Eq. 13 and 14:

$$(QF50P)^2 (Q7L2)=54.93 (QAA)^{3\cdot 00} \qquad ......(15)$$

$$QF50P=7.41\left[\frac{(QAA)}{(Q7L2)^{0\cdot 33}}\right]^{1\cdot 50} \qquad ......(16)$$

In general

$$QFX= C\left[\frac{QAA}{(Q7L2)^{s}}\right]^{m} \qquad ......(17)$$

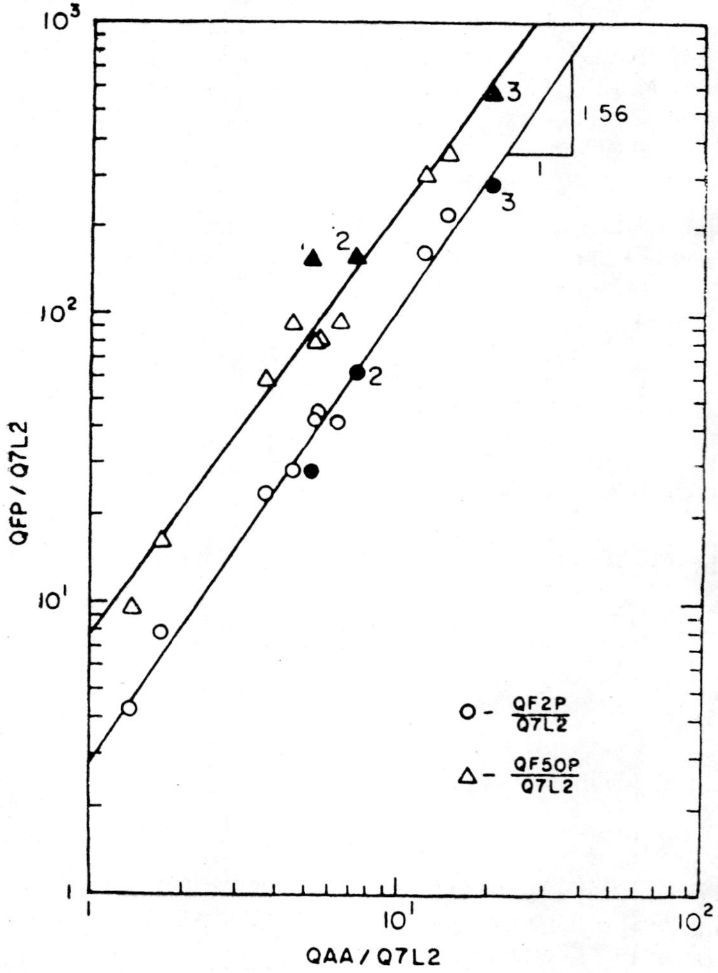

FIG. 5.8 : *Flood: Low flow ratios related to average low flow ratios.*

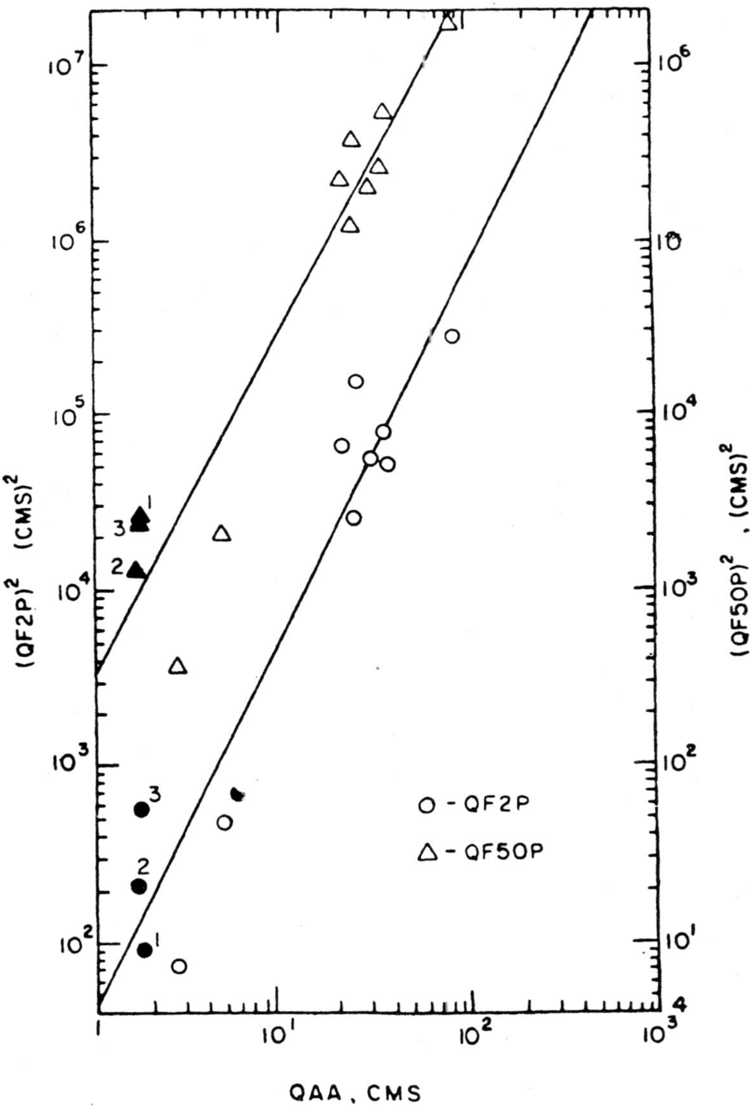

FIG. 5.9 : *Peak flood flows squared related to average flow*

Comparisons of the results of graphical solutions to the 2-yr. or 50-yr. equations in Figs. 5.4a and 5.4b (Eqs. 4 and 6), with the results of the semi-empirical analysis of the Lewis Study Area (Eqs. 14 and 16), are presented in Table 5.2. A different sample of Lewis Study Area gaging stations was used in each of these analyses, although both samples had some stations in common. Note the self-adjusting nature of the exponents (s) and (m) derived from the original $QF2$ relationship in Eq. 4. Their multiple is always equal to 0.5.

TABLE 5.2 : **Comparisons of coefficients and exponents derived by graphical and semi-empirical methods in the form of Eq. 17.**

Method (Eq. No.)	Flood Flow	Exponent s	Exponent m	Coefficient C
Graphical Regression				
(Eq. 4)	QF2P	0.32	1.54	2.88
(Eq. 6)	QF50P	0.34	1.47	7.87
Semi-Empirical				
(Eq. 14)	QF2P	0.33	1.50	3.18
(Eq. 16)	QF50P	0.33	1.50	7.41

The 50-yr. flood equations display considerably more variability than do the 2-yr. flood equations because very few gaging stations have long enough periods of record to accurately determine the 50-yr. floods. To test the validity of the analytical result that the exponent (m) in the 50-yr. flood equation is 1.50, a sample of long-term station records was analyzed. Eight stations with a minimum of 30 yrs. of record were used for semi-empirical analyses from Province 2-A on the west slope of the Cascade mountains in Washington. The resulting prediction equation is :

$$QF50P = 11.68 \left[\frac{(QAA)}{(Q7L2)^{0.33}} \right]^{1.53} \qquad \text{......(18)}$$

which is comparable to the results obtained using computerized regression analyses with many more data points. The results of the regression analysis for all data points in part 2-A of Washington Province 2 from which the data sample for Eq. 18 was derived, is :

$$QF50P = 8.98\left[\frac{QAA}{(Q7L2)^{0.33}}\right]^{1.53} \qquad \ldots\ldots(19)$$

Two methods of developing flood prediction equations which relate flood flows of two frequencies to average annual flows and low flows in a hydrologic province or smaller area have been presented. Both produce essentially identical results. Certainly when large volumes of data are available the computerized regression analyses is more satisfactory. For limited data, or if a quick first estimate is desired, the semi-empirical method performs satisfactorily.

The methods for determining the average annual and low flows in ungaged streams and portions of streams above gages are discussed in an earlier section. Once these flows have been determined, the 2-yr. and 50-yr. floods, either peak or maximum daily, for that ungaged portion of a stream can be determined. From these two flood flows a flood frequency (or recurrence interval) curve for the ungaged stream can be prepared. The estimated value of the 100-yr. flood flow (or higher) can be determined by extrapolation.

RESULTS OF THE PROJECT

Regression equations

The locations of the various provinces in the three states were shown in Fig. 5.1. The coefficient *C* and the exponents (s) and (m) used in this equation:

$$QFX = C\left[\frac{QAA}{(Q7L2)^{s}}\right]^{m} \qquad \ldots\ldots(17)$$

are summarized for all the provinces, and for 2-yr. and 50-yr. floods in Table 5.3. In Eq. 17, (*QFX*) is used to denote the flood flow of (*X*) recurrence interval (*X*-axis), either (*QF*2) or (*QF*50) and either the maximum daily average floods (*QF*2*D*) and (*QF*50*D*), or the peak floods (*QF*2*P*) and (*QF*50*P*). Because both types of flood frequencies are of interest, the Oregon data was analyzed both for floods based on the maximum annual one-day average flows and the annual instantaneous peak flows. The coefficient and exponents for these two types of floods are presented in the centre of two columns of Table 5.3. Washington provinces 4 and 5 were combined due to a lack of sufficient data in either province by itself.

TABLE 5.3 : **Results of low, average and flood flow analysis for Washington, Oregon and Idaho**

Washington				Oregon								Idaho			
2-Yr Peak Floods				2-Yr Avg. Daily Floods				2-Yr Peak Floods				2-Yr Peak Floods			
Prov	m	s	C	Prov	m	s	C	Prov	m	s	C	Prov	m	s	C
1	1.46	0.34	5.35	1	1.30	0.38	5.41*	1	1.09	0.46	14.78*	1	1.46	0.34	2.75
2-A	1.55	0.32	3.68	2	1.54	0.32	2.19	2	1.44	0.35	4.07	2	1.49	0.34	2.75
2-B	1.46	0.34	4.00	3	1.49	0.34	2.63	3	1.43	0.35	4.00	3	1.52	0.33	2.74
3	1.48	0.34	3.42	4	1.42	0.35	1.39	4	1.42	0.35	1.44	4	1.43	0.35	2.50
4-5	1.41	0.35	3.36	5	1.41	0.35	1.84	5	1.39	0.36	2.26	5	1.44	0.35	3.31
6	1.45	0.35	2.86	6	1.46	0.34	2.37	6	1.20	0.42	3.91				
8	1.38	0.36	4.04	7	1.48	0.34	1.98	7	1.45	0.35	2.45				
50-Yr Peak Floods				50-Yr Avg. Daily Floods				50-Yr Peak Floods				50-Yr Peak Floods			
Prov	m	s	C	Prov	m	s	C	Prov	m	s	C	Prov	m	s	C
1	1.40	0.36	14.00	1	1.15	0.44	17.43*	1	0.80	0.61	37.36*	1	1.38	0.36	61.40
2-A	1.53	0.33	8.98	2	1.42	0.35	7.10	2	1.48	0.34	8.49	2	1.39	0.36	9.41
2-B	1.39	0.36	12.42	3	1.46	0.34	6.57	3	1.44	0.35	9.96	3	1.46	0.34	6.73
3	1.43	0.35	8.82	4	1.14	0.44	3.58	4	1.15	0.44	3.71	4	1.27	0.39	10.21
4-5	1.25	0.40	10.50	5	1.42	0.35	5.20	5	1.42	0.35	6.59	5	1.25	0.40	6.20
6	1.25	0.40	8.55	6	1.48	0.34	9.27	6	1.18	0.43	19.03				
8	0.98	0.51	15.51	7	1.57	0.32	3.35	7	1.51	0.33	5.13				

Note: Equation 17 is the general form ; * only 4 data points.

The 50-yr. exponent values of (s) and (m) tend to be further away from 0.33 and 1.50 than the 2-yr. values because of short periods of record and relatively few data points in some provinces. In using the regression analyses, all points are weighted equally in the determination of (C), (s) and (m). Therefore, without common periods of record and records of considerable length, 50-yr. flood equations would be difficult to develop on a common basis for direct comparison.

The equations are, of course, very sensitive because of the large coefficient (m) and small exponent (s) involved. But the equations are very stable and reflect natural conditions from which the effects of man-made activities can be determined. In Idaho province 2, two separate analyses were conducted for different periods of record, one until 1966 and the other until 1972. The results were identical with the plotting points for the longer period merely raising the graphical relationships developed for the shorter period. This shift reflected a general increase in average annual flows and flood flows during the period 1967-72 compared to the shorter period.

The relative consistency in the 2-yr. exponents (s) and (m) values in most of the Washington, Oregon and Idaho provinces tends to indicate that the variability in the coefficient (C) could be correlated with some hydrologic or geographic factor. Several trial solutions were made in an attempt to combine all of the more provincial results into one equation, but none were adequately successful.

Semi-empirical analyses

As discussed earlier, the semi-empirical combined graphical and mathematical method of analyzing a selected sample of data points from a province has some distinct advantages over the regression solution for provincial flood equations.

Semi-empirical analyses were completed on samples of 2-yr. flood data for comparison with the regression equations in Washington provinces 2, 4-5, and 6; and in Idaho province 2. The comparative values of (C), (s) and (m) between the semi-empirical and regression analyses are presented in Table 5.4. Quite good agreement is evident between the less extensive semi-empirical analysis of provincial data samples (6 to 10 values) and the total data set (30 to 70 values).

TABLE 5.4 : **Comparative results of semi-empirical and regression analyses for 2-yr. floods in selected provinces in Washington, Oregon and Idaho**

State	Province	Semi-Empirical			Computer Regression		
		C	s	m	C	s	m
Washington (QF2P)	2*	5.15	0.35	1.44	3.85	0.33	1.50
	4-5	3.91	0.35	1.44	4.63	0.35	1.41
	6	2.21	0.33	1.50	3.42	0.35	1.45
Idaho (QF2P)	2	2.90	0.34	1.49	2.85	0.34	1.49

*Provinces 2A plus 2B.

APPLICATIONS OF RESULTS

Testing of prediction equations

Two stations in Washington, one in province 3 and one in province 8, were selected to test the prediction equations. The

TABLE 5.5 : **Estimation of two-year and fifty-year peak floods in Washington provinces 3 and 8.**

Factor	Unit	Salmon Cr. Prov. 3 (Sta. 14-2120)	Asotin Cr. Prov. 8 (Sta. 13-3345)
Drainage Area, A	sq km	47.40	404.00
Average Precipitation	cm/yr	147.32	71.12
Basin Relief, H	km	0.37	1.22
Recorded QAA	cms	1.73	1.94
Predicted QAA	cms	1.77	1.73
Recorded Q7L2*	cms	0.085	0.78
Predicted Q7L2*	cms	0.093	0.85
Predicted QF2P**	cms	22.92	9 37
Recorded QF2P	cms	23.97	9.28
Predictcd QF50P**	cms	59.43	28.58
Recorded QF50P	cms	48.68[a]	31,70[b]

*Diversions above both gaging stations

**Using computer regression analyses for provinces and predicted QAA and Q7L2 values

[a]Extrapolated based on 28 yrs. of record;

[b]Based on 44 yrs. of record

low flows (*Q7L2*) were estimated using the basin parameter methods described in Ref. (3). The average annual flows (*QAA*) were estimated using regional correlations of gaged flows with average annual precipitation and drainage area[3] ($QAA = C(P.A)$). Neither station was used in the provincial regression analyses. The results are tabulated in Table 5.5

CONCLUSIONS

There is a very strong basic correlation between low, average and flood flows such that :

$$Q7L2 = f\left[\frac{(QAA)^3}{(QF)^2}\right] \qquad \ldots\ldots(20)$$

In words, low flow (*Q7L2*) to the first power is a function of flood flow squared $(QF)^2$ and average annual flow cubed $(QAA)^3$. The coefficient (f) in Eq. 20 varies between hydrological provinces, but for natural flows and no severe geologic anomalies the coefficient is very consistent within provinces. It is smaller for arid regions and larger for small, steep watersheds at high altitudes having heavy snowpack, but relatively small low flows due to a lack of groundwater storage.

The concept for developing this model relating one flow to another flow depends on the basin as the integrator of all flow processes. Provincial and subprovincial relationships are generated using the best stream-flow records available, meaning records with the least regulation, least diversion, longest period of record and, thus, the most natural flow.

The relationships should be evaluated by regression analysis and by graphical inspection in order to test the homogeneity of the data. It is extremely important, especially for flood flows of larger recurrence intervals (i.e., *QF*50), that only those stations with the longest and most common periods of record be used in developing the correlations.

It has been shown that certain parts of the Pacific North-West Region (i.e., Idaho) have only a slight variation in the coefficient and powers in the flood prediction equations. For the 2-yr. flood (*QF* 2) the coefficient (*C*) had only a 10 per cent variation throughout the state of Idaho, and less than a 2 per cent variation in the powers (m) and (s). The other variations are due primarily to a paucity of record. Further investigation into the coefficients in the 1, 2, 3, power equation is under way.

REFERENCES

1. P., Bock, et al., 1972 : Estimating peak runoff rates from ungaged small rural watersheds, *Natl. Coop. Hgwy. Res. Program Report 136*, Highway Research Board.
2. John R., Crippen, Jan., 1974 : *Basin Characteristics Indexes as Flow Estimators*, ASCE National Meeting on Water Resources Engineering, Los Angeles, Calif.,
3. John F., Orsborn, N. Sood and Mohinder Oct., 1973 : *Technical Supplement to the Hydrographic Atlas*, *Lewis River Basin Study Area*, State of Washington, Dept. of Ecology, State Water Program.
4. Brian M., Reich, July, 1971 : *Runoff Estimates for Small Rural Watersheds*, Final Project Report to Federal Hgwy. Comm. for Contract FH-11-7429., The Pennsylvania State Univ.
5. D.M. Thomas, and Benson, M.A. 1970 : Generalization of streamflow characteristics form drainage-basin characteristics, *U.S. Geological Survey, Water Supply Paper 1975*.

SIX

V. Gardiner

DRAINAGE BASINS MORPHOMETRY
Quantitative Analysis of Drainage Basin Form

INTRODUCTION

THE TERM morphometry is used in several disciplines to mean the measurement and analysis of form characteristics; in geomorphology it is applied to numerical examination of landform, which may be more properly termed geomorphometry. This may be concerned with landform as a whole (general geomorphometry) or with a particular type of landform (specific geomorphometry). Whilst no data source is excluded by this definition most geomorphometric studies have been based upon map derived data. The drainage basin is the landform most commonly analysed in morphometry, and since 1940 geomorphological literature has included many hundreds of studies of drainage basin form. Fluvial morphometry has become a fundamental concern of much post-war geomorphological research in the United States and Great Britain, and is being increasingly widely used in other areas, especially India (Vaidyanadhan, 1977), Brazil (Christofoletti, 1969), Australia and Canada. This has resulted from, first, the recognition that the drainage basin is the fundamental unit of landform in fluvial terrain (Chorley, 1969); second, the need for quantitative information on morphology to accompany measurements of processes in process-response models; and third, the general change in the philosophy and methodology of geomorphology from a subjective and deductive science based upon observation to an objective and inductive science based upon measurement. The vast and burgeoning literature of drainage basin morphometry has previously been reviewed by Strahler (1964), Gregory and Walling (1973), Gardiner (1975 and in press), and Gardiner and Park (1978). The purposes of the present chapter are to outline the techniques used in drainage basin morphometry, to provide a brief account of the development of the subject, and to assess its present day status and potential.

Techniques

Following Gardiner (in press) it may be suggested that drainage basin morphometry consists of five stages, namely network delimitation, sampling, measurement, variable definition, and analysis. Although many authors advocate the use of field survey to identify the functioning drainage network this is impossible for large areas, and recourse has of necessity to be made to either cartographic or remote sensing sources (including aerial photographs). These have limitations as summarised in Table 6.1 (Gregory and Gardiner, 1979) and illustrated by Fig. 6.2. When deriving data from maps it is necessary to know of the surveyor's conventions, which are responsible for his definition of streams during field survey, and to be aware of the generalisation procedures which are involved in production of small-scale maps (Chorley and Dale, 1972). Different mapping agencies have evolved different procedures for definition and depiction of the river network (Drummond, 1974; Ovenden and Gregory, 1980). Field-derived data are also subject to variation because of temporal variability in stream network extension, and many studies (Fig. 6.1) have illustrated the way in which drainage density responds to changing meteorological conditions and, hence,

TABLE 6.1 : **Some limitations of operational procedures for delimitation and measurement of drainage networks.**

	Topographic map	Remote sensing imagery
Convention	Surveyors coventions for field mapping.	Type of imagery
Processing	Plotting of stream network for map of definitive scale.	Criteria for channel definition.
Scale	Generalisation of network at different scales.	Resolution of sensor
Date of Survey	Significant because of possible changes of conventions and processing, and temporal changes of network extent.	

discharge. Use of remote sensing sources is similarly dependent upon an awareness of the characteristics of stream channels on the types of imagery employed, of the criteria by which channels may be defined on such imagery, and of limitations imposed by the resolution of the sensor.

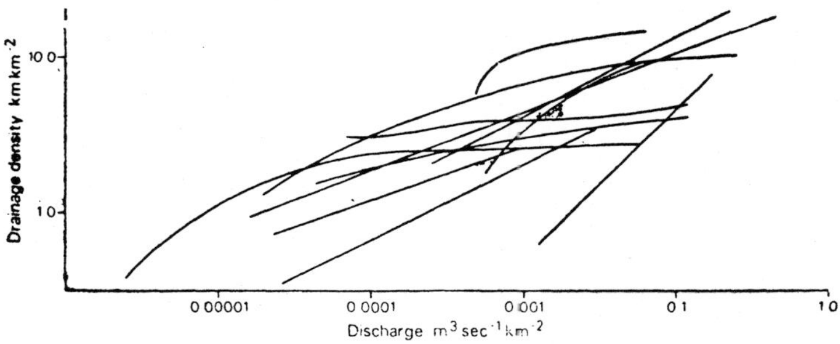

FIG. 6.1 : *Relationships between drainage density and discharge*

Sources : Dunne (1976), Gurnell (1978) and sources therein.

Whatever the source of information networks may be delimited in various ways (Table 6.2; Fig. 6.2). Whilst the blue lines depict the most obvious network to employ the inadequacies of stream network depiction on American maps led Morisawa (1957) to suggest the use of contour crenulations to identify channels which exist in the field but which are not depicted by blue lines. This procedure has, however, been criticised by Gregory (1966) in Britain because it includes in the network features which may have originated under former morphogenetic conditions, and which, therefore, represent fossil palaeohydrological elements. If the crenulation method is used it is necessary to employ a consistent rule to determine the lengths of exterior links in the network. Shreve (1974) suggested that the channel source should be defined as that point where the channel slope declines to a pre-defined value. Smart (1978) has successfully employed this, and Gardiner (in press) reviews several alternative procedures. A final basis for network delimitation is employed by the mesh-length extension network (Fig. 6.2; Table 6.2). This represents the maximum network, and is obtained by extending all exterior links in the network, defined from either blue lines or contour crenulations, headward to the watersheds. This method gives an over-estimate of the extent of the network actually existing, and is rarely used.

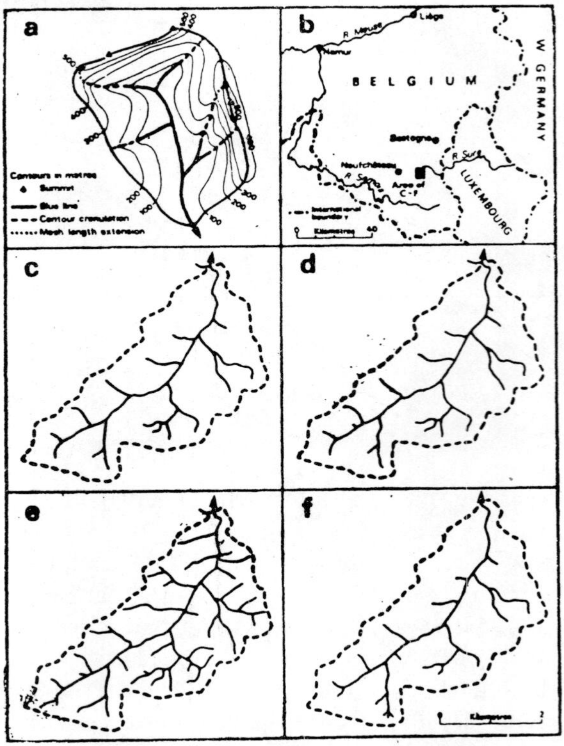

FIG. 6.2 : *Aspects of network delimitation*

1a. Alternative networks (See Table 6.2.).
1b. Study area of 1c-1f and Figures 6.4 and 6.5
1c. Drainage network from EGM 1:50000 map
1d. Drainage network from EGM 1:25000 map
1e. Contour crenulation defined network from EGM 1:25000 map
1f. Network defined by morphological criteria in field.

TABLE 6.2: **Bases for delimitation of drainage networks**

Basis	Map (letters refer to Figure 6.2)	Field
Hydrological	(a) Blue lines. Governed by cartographic conventions	Flow during survey. Temporally variable
Morphological	(b) Contour crenulation network. Consistent termination rules necessary.	All channels. Consistent rules for identification necessary.
Maximum	(c) Mesh length extension. Rarely used outside U.S.A.	Channel extensions into flushes, pipes and hollows. Difficult to implement.

Identification of a sample of drainage basins for which analysis may be carried out is commonly affected by some method of network ordering, unless there are *a priori* reasons for choosing an arbitrary sub-division of the entire network, as for example in studies concerned with relationships between morphometric properties and discharge as measured at existing gauging stations. Many methods of stream ordering have been devised (Fig. 6.3). Those of Strahler (1952) and Shreve (1966) have been most widely employed in morphometric investigations, and preliminary suggestions may be made concerning the most appropriate method for particular purposes (Table 6.3). Problems concerned with selection and use of an ordering method have been circumvented by recent computer based methods in which data are collected for individual links in the stream network and aggregated as required. Another device has been the use of regular units. Gardiner (1977) explored the use of grid squares for mapping drainage density, and relationships have also been identified between topological characteristics of networks in sampling quadrats (Dacey and Krumbein, 1976; Gardiner, 1979).

The measurement of lengths and areas is essential to virtually all morphometric investigations. Available manual methods have been reviewed by Gardiner (1975: in press) who suggested that the polar planimeter offered the most effective

TABLE 6.3 : **Characteristics of some commonly used network ordering methods**

Method	Horton 1945	Strahler 1952	Shreve 1966	Woldenberg 1967	Gregory and Walling 1973 and Graf 1975	Smart 1978
Symbol	—	—	m —magnitude	w	g	t
Derivation	—	see Figure 3	No. of exterior links (i.e. sources)	$w=\frac{\log m+1}{\log r}$ r=bifurcation ratio	g=2m−1	Each order has range of $4t^{-1}$
Properties (kep below)						
1		*				
2		*	*	*	*	*
3	*	*	*		*	*
4			*	*	*	*
5			*	*	*	
6		*	*		*	*
Suggested purposes	None	S	T/S	T	P	T/S

Key. Properties

1. —Is completely hierarchical, each basin of order K having at least one of order K−1
2. —Is completely objective in application
3. —Produces integer orders
4. —Does not violate the distributive law
5. —A change of order occurs at each junction
6. —Produces only one defined ordering for a given network

Purposes

S—Sampling

T—Topological investigations

P—Process studies

method of areal measurement and that the opisometer (map measurer) is most appropriate for linear measurements. Effectiveness in this context implies the best compromise between speed, cost, accuracy and ease of use. Digitisation by use of a co-ordinate digitiser linked to an electronic computer is an increasingly attractive and flexible alternative (e.g. Jarvis, 1976a), but the limited availability and low speed of such equipment has so far restricted its use. Investigations of network topology have, however, made extensive use of network representation in binary form as pioneered by Shreve (1967), and algorithms have been devised for extraction of network information from binary representations of networks (Smart, 1970, 1978; Ferguson, 1977).

For some purposes it is more desirable to examine large areas than to obtain extremely accurate data. In these cases it may be desirable to estimate values of morphometric variables using a relatively quick method. For example Mark (1974) concluded that although the line intersection method for drainage density estimation embraced an average error of about 10 per cent this could be counter-balanced by a 50 per cent saving in time, Gardiner (1979) investigated a topological method by which grid square drainage density could be estimated more quickly than by the line intersection method, and Richards (1979) has independently investigated its application to drainage basins. Methods have also been devised for rapid estimation of other indices; of particular significance is the effective procedure of Pike and Wilson (1971) by which the geomorphologically significant but operationally tedious hypsometric integral (Strahler, 1952) may be estimated.

Basic basin dimensions may be combined into a wide variety variables expressing significant elements of basin form, and the nature of the elements which it is wished to measure must necessarily influence the nature of the measurements made. The introduction of the plethora of variables proposed has been conditioned by a variety of purposes, scales and environments of individual studies. However, it has been suggested (Gardiner, 1978; Ebisemiju, 1979) that the range of variables available essentially represents only a few fundamental and conceptually irreducible elements of basin form. Thus, basin scale, gradient and shape, network geometry and extent appear to represent separate, if not necessarily independent, components of over all basin form. Some conclusions concerning the variables which may be used to express these components are summarised in Table 6.4, after Gardiner (in press).

For expressing scale basin planimetric area is usually used although basin perimeter is easier to measure. Order and magnitude are also convenient topological indices of scale, and scale has been shown to pervade many topological indices (Onesti and Miller, 1978).

Relief ratio has been widely accepted as an effective measure of gradient aspects of the basin, despite uncertainties surrounding definition of its component measurements (Table 6.4). For example at least ten alternative definitions of basin length have been proposed (Gardiner, 1975; Cannon, 1976), and basin relief may be unduly influenced by one isolated peak within the basin. The hypsometric integral offers a conceptually attractive alternative index, despite criticisms made of it (Aronivici, 1966; Harlin, 1978).

Basin shape may be indexed by simple dimensionless ratios of the basic measurements of area, perimeter and length (e.g. Singh, 1979), as reviewed by Gardiner (1975). A more sophisticated approach is that described by Jarvis (in press) and evaluated by McArthur and Ehrlich (1977), in which the basin outline is modelled by a Fourier function, the harmonics of which describe the form (Jarvis, 1976b). Anderson (1973) has developed a similar method for examining the three-dimensional form of the basin. The method of Blair and Biss (1967) may offer considerable potential for process estimation based upon shape measurements, since it is based upon the distribution of area within the basin.

Network extent is of considerable hydrological and geomorphological significance since it is an index which measures both the nature of dissection within the basin and the extent of the hydrologically-significant network. The variable most commonly used to measure this characteristic is drainage density, although others have been suggested (Table 6.5). Some alternatives have been evaluated by Singh (1976). For process prediction Gregory (1977) has argued that network extent should incorporate some allowance for channel volume, and has suggested how both channel cross-sectional area and

TABLE 6.4 : **Some commonly used variables for aspects of basin form.**

Basic measurements

Measurement	*Dimensions*	*Symbol*
Area	1^2	A
Perimeter	1	P
Total channel length	1	Ct
Basin length	1	L
Basin relief	1	Hb
Nos. 1st.-nth. order streams	—	$N_1 \ldots N_n$

Variables	*Derivation*	*Comments*
Basin scale		
Area	A	Strictly, is only planimetric equivalent of true area : difference negligible except for very steep basins.
Perimeter	P	Often difficult to delimit precisely.
Order/Magnitude	See Table 3	Topological analogue of scale.
Gradient and relief form		
Relief ratio	$R = Hb/L$	May be unduly influenced by one point. Length variously defined. (See Gardiner, 1975 and Cannon, 1976)
Hypsometric integral	See Strahler (1952)	Best determined by approximation (Pike and Wilson, 1971). May be area dependent (see Aronovici, 1966).
Basin shape		
Elongation	$Se = A^{0.5}/L$	Simple ratios—others also possible—see Gardiner (1975).
Relative perimeter Crenulation	$Sc = P^2/A$	
See text and Gardiner (1975) for other measures.		
Network extent		
Drainage density	$Dd = Ct/A$	Widely used : estimation method available—see text.
See also Table 5.		
Network geometry		
Stream frequency	$F = (N_1 + N_1 - 1)/A$	Related to D—see Melton (1958b)

Bifurcation ratio	Variously derived—see Haggett and Chorley, 1969	Recognized to be of limited utility
Magnitude	Greatest no. of links from mouth to source.	Topological analogue of mainstream length.

For others see Smart (1978) and Jarvis (1977a).

stream length may be incorporated into an index of network volume. This he subsequently refined (Gregory, 1979a) in the form of an index incorporating relief, and termed network power.

Available indices of network geometry are examined below (Table 6.5) and in Table 6.4. The simplest measures are those of bifurcation ratio and stream frequency. More sophisticated topological variables are reviewed by Jarvis (1977a) and Smart (1978).

The inherent variability of morphometric data almost inevitably requires some form of statistical analysis in order to identify fundamental relationships amongst the data. However, the nature of morphometric data is such that statistical analysis is attended by several groups of problems. First, large volumes of data are necessary in order to identify meaningful relationships, because of the variability inherent in morphometric data. This, therefore, requires the use of an electronic computer for computations. Secondly, the distributional characteristics of morphometric data are such that it is rarely appropriate to use parametric statistical procedures (Gardiner and Gardiner, 1979) without transformation of the data. For example Gardiner (1973) examined data for over 2000 drainage basins and suggested the most appropriate transformation of each of 23 variables. Similar conclusions have been reached elsewhere (e.g. Tandon, 1971; Shukla and Verma, 1974). Thirdly, the use of some statistical procedures demands a knowledge of the error variance incorporated into the data because of measurement and associated processes (Mark and Church, 1977). This knowledge is rarely available in morphometric studies, and is worth further attention. Finally, there is increasingly being demonstrated a need for a greater understanding of the meaning of correlation between variables (Carson, 1966).

TABLE 6.5 : **Some variables for expressing aspects of network extent and related concepts.**

Index	*Source*	*Derivation*	*Dimensions*
Drainage density	Horton 1932	$D=L/A$	Km^{-1}
Length of overland flow.	Horton 1945	$LF=1/2D$*	Km
Constant of channel maintenance.	Schumm 1956	$C=1/D$	Km
Stream frequency	Horton 1945	$F=N/A$	No. Km^{-2}
Relative stream density.	Melton 1958B	$R=F/D^2$	Dimensionless
Drainage intensity	Faniran 1968	$I=D\times F$	No. Km^{-3}
Network volume	Gregory 1977	$V=\int_o^T CdLe$	Km^3
Texture ratio	Smith 1950	Nc/P	No. Km^{-1}
Texture	Morgan 1976	Ns/A	No. Km^{-2}
Microscopic and macroscopic link density.	Smart 1972, Abrahams 1980	$\Phi=L^2/A$ $K=\bar{L}^2/\bar{A}$	Dimensionless

Ns=No. of stream sources
L=stream length
A=basin area
N=No. of streams of given order
Le=length of channel above cross-section measured
T=total length of channel in network
D=Channel cross-sectional area
Nc=No. of crenulations in most crenulated contour
P=Basin perimeter

*assuming mean channel slope $<1/3$rd mean ground slope

Morphometric data often consist of variables composed of various combinations of only a few basic measurements. This introduces the possibility that an apparently significant correlation, as judged against a null correlation of zero, is in fact a correlation between related ratios, which should be judged against a non-zero null correlation (Chayes, 1971). Thus, the nature of the relationship between drainage density (L/A) and basin area (A) has been hotly debated (Gardiner et al, 1977; 1978; Pethick, 1974; Ferguson, 1978). The redundancy inherent in the batteries of morphometric variables commonly employed has been shown to be considerable. For example Gardiner (1978) reduced 23 variables to only 5 components whilst retaining 88 per cent of the total variance in the data, and Onesti and Miller (1974) presented a similar study incorporating both basin and channel characteristics.

Analysis of morphometric characteristics may also be performed by non-statistical methods, and examination of the spatial variation of morphometric parameters is of considerable importance (Gregory, 1978). For example Mather and Doornkamp (1970) examined the spatial pattern of drainage basin characteristics in Uganda. Similarly Gardiner (1974) has mapped a number of drainage basin attributes in Devon, and has attempted to relate these to environmental characteristics (Gardiner, 1978) and to develop classification of land based upon them (Gardiner, 1976).

The preceding paragraphs have outlined the methods which are available for collection and analysis of morphometric data. A fuller appreciation of the ways in which such analyses may contribute to geomorphological investigations may be gained from a consideration of the development and present-day status of drainage basin morphometry. The development of this field of study has occupied five separate but nevertheless chronologically overlapping phases which Gardiner and Park (1978) recognised as the early antecedents, the formulation of basic concepts, intensification of investigation, evaluation and re-assessment of data and relationships, and the development of new directions.

THE DEVELOPMENT OF DRAINAGE BASIN MORPHOMETRY

Drainage basin morphometry was subject to three stimuli before 1940. Early hydrological ideas from Europe were imported into the United States by an American hydrologist, R, E. Horton, and in a series of contributions (Horton, 1926, 1932) he suggested how river discharge could be related to basin physiographic factors and how these factors could be indexed and measured. A second stimulus came from pedological and agricultural studies of the processes of runoff and soil erosion which were carried out in the 1920s and 1930s. The third stimulus was the increasing use and acceptance of numerical techniques in earth sciences in general and landform analysis in particular. Thus, by 1944 Neueschwander was able to prepare a 563 item review of early studies of slope and area-altitude relationships. Many of these are reviewed by Horton's seminal 1932 paper, which also describes some early and little-known attempts at fluvial morphometry.

These early threads were first brought together by Horton in 1932 and then further developed in his posthumous 1945 publication following verbal presentation in 1940. This remarkable and still widely-cited paper provided the concept of stream ordering which according to Bowden and Wallis (1964) is "the touchstone by which drainage net characteristics could be related to each other and to hydrologic and erosional processes." Ordering affords an objective procedure by which a complex stream network may be divided into constituent basins and by which the hierarchical structure or drainage composition of the network may be examined. Horton used ordering as part of an integrated approach to drainage basins, incorporating both their functioning and evolution in a type of process-orientated approach which was to fill the philosophical gap in geomorphology created by the recognition of serious weaknesses in the Davisian model of landscape genesis. It furthermore helped to unite work from the disparate disciplines of hydrology, geomorphology and, to a lesser extent, hydraulics. Horton explicitly formulated two basic 'laws' of drainage composition and a further supplementary 'law', all of which followed a geometric form relating the number of streams of each order, and their mean lengths and slopes, to the stream order, as illustrated in Fig. 6.4.

Approximately the first two-thirds of the period from 1945 to the present may be envisaged as encompassing a phase of post-Hortonian intensification and development, in which considerable effort was devoted to testing the relationships postulated by Horton in a wide range of environments and developing the methods employed. Much of this work was carried out in the United States under the inspiration of A.N. Strahler at Columbia (Strahler, 1964), and in the United States Geological Survey, following work by W.B. Langbein (Langbein, 1947). Initially work was directed towards establishing the general applicability of Hortonian relationships, often employing large amounts of data. For example Milling and Tuttle (1964) examined data for 9078 stream segments in

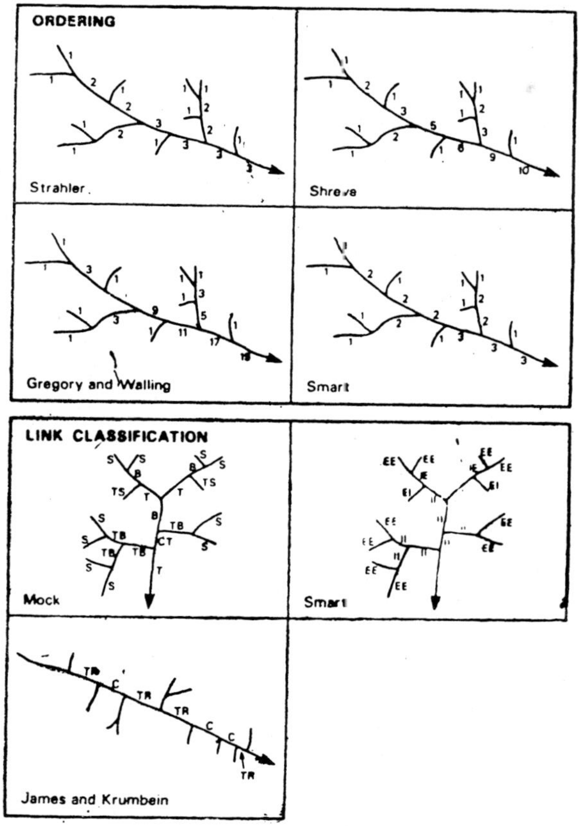

FIG. 6.3 : *Ordering and link classification methods. Link classifications are* :

S—source link
B—bifurcating link
T—tributory link
CT—cis-trans link
TS—tributary source link
TB—tributary bifurcating link
TR—trans link
C—cis link
EE—exterior link joining exterior link
EI—exterior link joining interior link
IE—interior link joining exterior link
II—interior link joining interior link

order to conclude that their drainage basins were 'normal'. This and similar studies in the United States were largely published in journals and reports of limited international circulation, and later similar work appeared elsewhere (e.g. Singh and Srivastava, 1974; Asthana, 1975). However, during the 1950s R.J. Chorley published a series of papers in internationally circulated journals (Chorley, 1957a, 1957b, 1958) which not only further developed and examined the techniques available but also brought them to the attention of geomorphologists outside the United States.

A wealth of further indices of drainage basin form have been proposed since Horton's initial examination of a small number of basic basin attributes. For example Strahler and his co-workers introduced indices such as relief ratio (Schumm, 1956) and ruggedness number (Strahler, 1958) which have since been widely employed. Other workers have concentrated on improving the operational definitions employed. For example Gardiner (1975) and Cannon (1976) have identified over ten different definitions of basin length, and aspects of drainage basin shape have also been the subject of many different indices, as reviewed by Gardiner (1975; in press).

Stream ordering techniques have also been the subject of much innovation. The original Hortonian method contained an element of subjectivity in its application. In 1952 Strahler introduced a modification which made it completely objective (Fig. 6.3) and this method is that which is still most commonly used. Bowden and Wallis (1964) have documented the effects of this change on the laws of drainage composition. In 1966 Shreve introduced a method of stream magnitude determination which took into account all links in the network, and other methods have since been developed; some of these are illustrated by Fig. 6.3 and reviewed by Gardiner (in press).

The basic set of three Hortonian 'laws' of network composition have also been supplemented by the discovery of further empirical relationships. For example in an important study published in 1956 Schumm introduced the laws of drainage basin areas and contributing areas and Fok (1971) has introduced one concerning stream relief. There has also been discussion concerning relationships between mainstream length and drainage basin area (Hack, 1957; Smart and

Surkan, 1967; Shreve 1974), and between link-length ratio and stream order (Ghosh and Scheidegger, 1970). These are illustrated by Fig. 6.5.

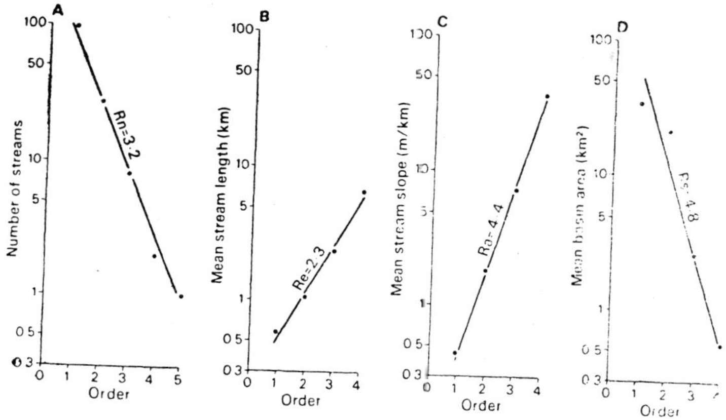

FIG. 6.4 : *Hortonian relationships for the headwaters of the R, Sure, Belgium, containing the small basin illustrated in Figure 6.2.*

The intensification phase of activity has encompassed the application of basin morphometry to a wide range of climatic and lithological environments. For example in India much information on the arid zones has been made available (Ghose et al., 1967; Ghose and Pandey, 1963; Singh and Ghose, 1973). Some synthesis of the many studies in different climates have been attempted by Gregory and Gardiner (1975) and Gregory (1976a). Individual lithologies studied have included pumice (Selby, 1968) and deeply weathered bedrock (Cunha et al. 1975), whilst comparison of morphometry between different rock types has been a common theme (e.g. Gardiner, 1971; Brunsden, 1969; Tandon, 1974; Padmaja, 1975; Sharma and Padmaja, 1977; Mithal et al., 1974). Dry valley systems have been analysed by the same methods (e.g. Gregory, 1976b; Morgan, 1971) and Karst drainage basins have been analysed by modifications to them (e.g. Bassett and Ruhe, 1973; Williams, 1972). Hortonian methods of analysis have even been applied to nonfluvial landforms such as glacial troughs (Haynes, 1972) and sand dunes (Goudie, 1969).

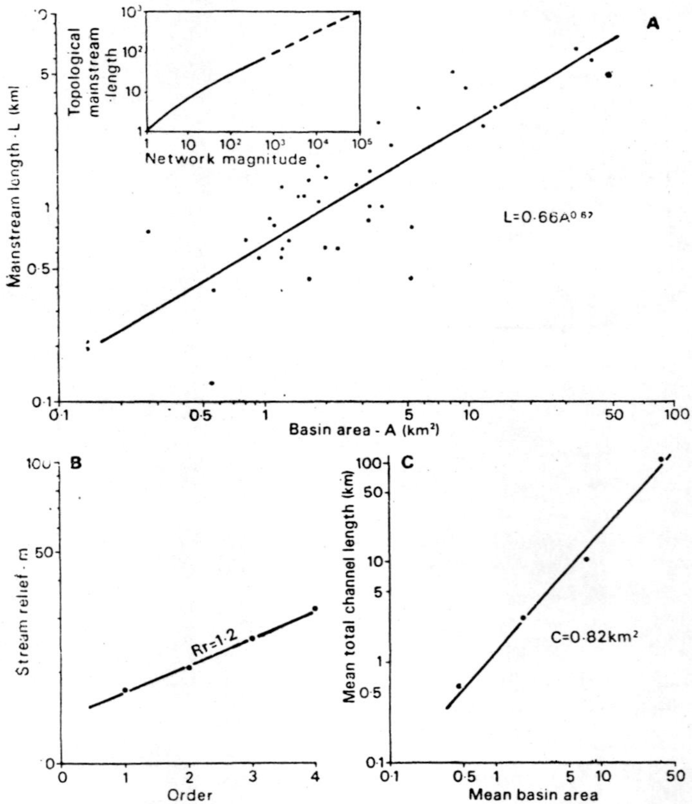

FIG. 6.5 : *Additional Hortonian-type relationships for the same basin as Figure 6.4*

The inset in Fig. 6.5a shows Shreve's theoretical derivation of topological mainstream length/network magnitude relationship, assuming topologically random channel networks. The slope is near 0.6 for magnitudes less than 100, and decreases towards 0.5 for larger basins, in a manner analogous to Hack's mainstream length/basin area relationship.

A final area in which intensification and development of Hortonian methods has occurred is that many attempts have been made to render the measurement of morphometric data less tedious, error-prone and time consuming. For example hypsometric integral (Strahler, 1952) may be effectively estimated from a simple method described by Pike and Wilson (1971). The practical and conceptual difficulties involved in determination of drainage density may also be circumvented by a variety of methods reviewed by Gardiner (1979), who concluded that a topological method was most effective in certain cases.

Since the mid-1950s the Hortonian type of analysis has been subject to much critical re-evaluation with respect to the limitations inherent in the data sources and to the interpretation of the relationships identified. There has been a realization that the cartographic data sources used in most morphometric studies are not necessarily a true portrayal of the drainage network in the real world (Chorley and Dale, 1972). Inconsistencies between mapped drainage networks and those occurring in the field have been identified in many areas, as reviewed by Gardiner (in press), and it is usually accepted that the veracity of mapped stream networks must be checked by field examination (e.g. Werrity, 1972) or aerial photography (e.g. Eyles, 1966). The problem of inadequate depiction of stream networks on maps is made worse by the recognition that stream networks vary in extent temporally in response to external conditions such as precipitation input. The significance of this for morphometric analysis is examined by Gregory and Gardiner (1979) and Gurnell (1978) who summarize available findings (Fig. 6.1).

Re-examination of earlier concepts has also been occasioned by the proliferation of morphometric indices proposed during the subject's intensification phase. Melton (1958a, 1958b) suggested how variables describing aspects of basin morphometry might be related to one another, and Strahler (1958) and Wong (1975) have attempted to develop small sets of independent dimensionless descriptors of basin form. It now appears that relatively few fundamental aspects of basin form are described by the many indices available (Gardiner, 1974, 1978), and that it is possible to envisage 'reduced-rank' models of basin form (Ebisemiju, 1979) in which much of the redundancy inherent in the use of multiple indices is removed.

A final, and probably the most important, way in which basin morphometry has been re-assessed is that concerning the fundamental meaning of the 'laws' of drainage composition formulated by Horton and others. Early interpretations envisaged them as empirical generalisations which applied to all river networks and which possessed fundamental geomorphological significance. Eventually it was realised that the 'laws' were in reality statistical generalisations which must inevitably apply to any branching phenomena and which result from the rules of the ordering system adopted. Milton (1966) demonstrated the geomorphological irrelevance of the 'laws' by applying them to a plum tree, and whilst this may have been an overstatement of the case (Robinson, 1967) analogies to biological and other similar networks (Leopold, 1971; Woldenberg and Berry, 1967) have emphasised that the law of stream numbers at least is nothing more than a statistical probability function which is conditioned by the use of a particular ordering system. Perhaps paradoxically this realisation has greatly aided the elucidation of regularities in drainage network composition, for whilst it seems a negative conclusion in the context of the initial Hortonian formulation it has stimulated fuller consideration of river networks as branching phenomena controlled by random topological processes (Shreve, 1975). Development of this approach is examined further below.

The decade of the 1970s has witnessed application of drainage basin morphometry in ways not initially envisaged. These new directions are characterised by both new approaches and new purposes of investigation. The major new approach developed amounts to no less than a new model of drainage network composition which has supplanted Horton's laws and which is known as the random model of drainage composition. This developed from the topological approach to drainage networks inherent in much of the re-evaluation phase of development discussed above. In Hortonian network analysis attention is usually focused upon properties of Strahler-order defined streams or stream segments or basins. In the random model the individual stream link is examined as the basic unit of the network. Three kinds of geomorphological variables may be associated with each link (Table 6.6), namely link, path and network variables (Smart, 1978). Link variables are associated with an individual link, path variable with the path from the link to the outlet, and network variables with the sub-network defined by and above the individual link. These are more closely associated with geomorphological properties than are Hortonian variables (Werner and Smart, 1973; Jarvis and Werrity, 1975).

TABLE 6.6 : **Link-associated variables**

Type of variable	Examples
Link	Length Associated drainage area Orientation
Path	Topologic path length Flowpath length Sinuosity
Network	Magnitude Basin area Total channel length Drainage density Link frequency

The random model of drainage composition is a quantitative formulation of general ideas originally put forward by Leopold and Langbein (1962). It was initiated by Shreve (1966) and has subsequently been extensively tested and developed, as reviewed by Shreve (1975) Smart (1978) and Jarvis (1977a). It can be summarised in two basic postulates after Smart (1973a), as :

1. In the absence of geological controls natural channel networks are topologically random.
2. The exterior and interior link-lengths of drainage networks developed in similar environments are independent random variables with a single common distribution for each type.

A third postulate concerning areal properties of interior and exterior link drainage areas has also been partially developed (Shreve, 1974; Smart, 1976; Smart and Werner, 1976), but has yet to be stated explicitly with a complete theoretical derivation. A fourth postulate concerning channel slopes has also been suggested in outline form (Flint, 1976). These postulates are generally found to apply to observations from natural maturely-developed drainage basins, although with exceptions for certain specific areas or details. Horton's laws can be explained as necessary outcomes of the model and the model can be used to estimate values of geomorphological relationships, as demonstrated by Smart and Werner (1976) and Smart (1978). The initial model does, however, have to be modified in order to examine certain network properties. For example James and Krumbein (1969) sub-divided interior links into cis and trans links (Figure 6.3) and showed how the proportions of these in a sample of observed main channels were not adequately predicted by the random model. Smart and Wallis (1971) explained the observed discrepancy as a result of spatial non-randomness imposed by headward network development and Flint (1980) has examined tributary asymmetry further. Further sub-divisions of link type have been proposed by Mock (1971) and Smart (1978), Figure 6.3, but the frequencies of occurrence of these have only been partially investigated by for example Abrahams and Campbell (1976).

Another way in which the model has been shown to be less than a complete explanation of network composition is that there have been suggestions that the composition of drainage networks is necessarily scale dependent. Thus, Jarvis (1977b) has made a distinction between order-formative basins, which characterise the headwaters, and order-excess basins which mainly occur in the lower parts. He has also shown (Jarvis, 1976a) how stream link-length organisation is network scale dependent and is sensitive to spatial scale threshold constraints.

The topological approach, based on the random model of drainage composition, has increasingly moved emphasis away from the basin *per se* to the network. A reaction to this has, however, occurred recently, in the from of approaches which place attention on the ridge network (Werner, 1972, 1973; Mark, 1979), rather than that of the streams. At a higher level of abstraction attempts have also been made to consider works by reference to point patterns formed by stream sources or junctions (Oeppen and Ongley, 1975; Dacey and Krumbein, 1971).

The second major dimension of innovation in drainage basin morphometry has been its application to both the prediction of processes and to other purposes. Basin morphometry was originally formulated with the aim of predicting water discharge for ungauged basins (e.g. Boulton, 1965; Langbein, 1947), and such attempts have continued (e.g. Patton and Baker, 1976; Gupta et al., 1969). A number of morphometric indices have been used in such studies, including drainage density. Whilst many early and some later attempts followed an entirely static 'black-box' approach in which drainage density has one value for each basin the realization that drainage density is a dynamic quantity which demonstrates temporal variation in sympathy with that of discharge (e.g. Gurnell, 1978) and controlled by vegetation and antecedent conditions has led to re-examination and re-interpretation of many earlier relationships (e.g. Gregory and Walling, 1968; Gregory and Gardiner, 1979). Surprisingly little success has resulted from attempts to relate stream discharge to traditional indices of basin morphometry, although a comprehensive study (N.E.R.C., 1975) has reported some encouraging results for floods. This has stimulated the development of further indices, such as those of network power, which Gregory (1979a) has shown to be an effective predictor of mean annual flood. This index combines both network and channel characteristics in one measurement.

Attempts have also been made to relate runoff to purely topological properties of the network, in both empirical (e.g. Surkan, 1974; Delleur and Lee, 1973) and theoretical (e.g. Kirkby, 1976) studies.

Sediment yield has been related to watershed characteristics, following early work in the United States such as that of Anderson (1957), Maner (1958) and Schumm (1956). More recent studies have been concerned with Australia (Abrahams, 1972c) and the tropics (Douglas, 1968), and attempts have also been made to examine the impact of aspects of basin morphology on solute production (Kirchner, 1965).

Other purposes for which drainage basin morphometry has been employed include the measurement of land form in a general sense, and in the formulation of models of landscape development. It has been suggested (Woodyer and Brookfield, 1966) that, since stream processes and consequently basin form are conditioned by all environmental characteristics which influence the hydrological cycle, basin form can be used to delimit physical regions in an objective manner which incorporates all significant environmental information. Thomas (1969, 1976), Singh (1971) and King (1970, 1975) have employed basin-derived indices in descriptions of terrain and Gardiner (1971, 1974, 1976, 1977) has examined the spatial variation of morphometric indices and used them in land classification. Morphometric information has also been incorporated into landscape inventories (e.g. Proctor, 1969; Bedient et al., 1977).

Basin morphometry has, finally, been employed in the formulation of models of landscape evolution and in the establishment of degrees of internal adjustment (e.g. Kumar and Pandey, 1975; Singh, 1976; Agrawal, 1972; Singh and Kumar, 1969). Ouma (1970) suggested patterns of variation of drainage basin measurements through time, and a number of early studies of morphometric development (e.g. Schumm, 1956; Melton, 1958a) have been re-examined (Abrahams, 1970) and alternatives proposed (Abrahams, 1972b, 1975, 1976; Wilcock, 1975). Following Ruhe's (1952) pioneer study temporal changes in networks have been examined by means of the ergodic hypothesis whereby spatial variation is viewed as an analogue of variation through time, despite debate concerning the assumptions made (Abrahams, 1972a, 1973; Smart, 1973b). Examination of morphometric inter-relationships has extended to examination of relationships between basin form and valley side slope form (e.g. Verma, 1972) and Richards (1977) has demonstrated how earlier apparently contradictory findings concerning valley side slope and basal stream gradient may be reconciled by a morphometric model in which the nature of the relationship is controlled by the nature of the slope processes operating. Prasad and Verma (1975) have similarly examined relationships between basin morphometry, slope form and slope processes.

CONCLUSIONS

Drainage basin morphometry is potentially a most important approach to geomorphology, since it affords quantitative information on large-scale fluvial landforms, which make up the vast majority of the earth's surface. However, the euphoria regarding the potential development of fully integrated drainage basin models which was characteristic of work in the 1960s has now largely disappeared. Recent work has proceeded in an atmosphere of more realistic caution and pragmatism, conditioned by the recognition of limitations imposed by the many theoretical and practical considerations reviewed above by the lack of a universally acceptable solution to the scale problem in geomorphology, and by a recognition of the complexities of process-response relationships which may exist (e.g. Brunsden and Thornes, 1979). For example not only may the spatial extent of the network vary temporally but the pattern of dependency amongst the controlling variables may also vary through time (Schumm and Lichty, 1965; Schumm, 1977).

There are, however, several very promising directions and emphases evident in many recent contributions to drainage basin morphometry. First, there appears to be an increasing willingness to envisage nominally separate levels of fluvial organisation as being in reality inextricably related. Whereas studies have in the past tended to be concerned with either river channels or basins it is now becoming increasingly common to find studies employing concepts and information from both levels. For example Onesti and Miller (1974) examined both network and channel properties in a study of patterns of fluvial variation, and Park (1976) has employed aspects of basin morphometry to explain variations in channel geometry similarly Gregory (1976c) has indicated the significance of channel typology to an understanding of fluvial networks, and has extended this to a preliminary examination of how channel typology may aid in elucidation of palaeohydrological relationships (1979b). However, despite these attempts the construction of fully integrated process-response models remains an elusive goal.

A second encouraging emphasis of much recent work has been the use of increasingly sophisticated statistical and mathematical models and techniques for data aquisition and analysis. Remote sensing data sources (e.g.Gustafson, 1973; Singh et al., 1969; Verstappen, 1977) will undoubtedly contribute much significant data in the future, from a wider variety of environments than has hitherto been possible. Acquisition and analysis of morphometric data is also increasingly being performed by electronic digitizer and computer. In addition the use of statistical methods for their own sake, which characterised some work of the 1960s and 1970s, is now tempered by a greater understanding and awareness of the assumptions inherent in the statistical procedures employed (e.g. Mark and Church, 1977), by the properties of morphometric data (Gardiner, 1973; Tandon, 1971) and by a somewhat reduced likelihood of techniques being employed merely because of their fashionability (Jennings, 1973).

A final encouraging factor which emerges from a consideration of recent work is that drainage basin morphometry appears to be the concern of a wide variety of researchers, in many geographical regions and with interests in a wide variety of geomorphological and non-geomorphological problems. Progress would, therefore, seem to be assured, providing that individual workers take care to recognise the contributions of related work and of work elsewhere. However, Gardiner and Park (1978) recognised that a dichotomy inevitably exists between the specialisation which individual workers must necessarily develop in their own subjects, and the generalisation which is necessary if one is to envisage a satisfactory integrated process-response model incorporating elements of basin morphometry. Undoubtedly it is the extent to which this dichotomy is reconciled that will control the future development and well-being of drainage basin morphometry.

SUGGESTED READINGS

Abrahams, A D., 1970 : An evaluation of Melton's order-by-order growth analysis, *Australian Geographical Studies*, 8, 57-70.

Abrahams, A.D., 1972a : Environmental constraints on the substitution of space for time in the study of natural networks, *Bulletin of the Geological Society of America*, 83, 1523-30.

Abrahams, A.D., 1972b : Factor analysis of drainage basin properties : Evidence for stream abstraction accompanying degradation of relief, *Water Resources Research 8*, 624-33.

Abrahams, A.D., 1972c : Drainage densities and sediment yields in eastern Australia, *Australian Geographical Studies*, 10, 19-41.

Abrahams, A.D., 1973 : Environmental constraints on the substitution of space for time in the study of natural channel networks : Reply. *Bulletin of the Geological Society of America*, 84, 353-6.

Abrahams, A.D., 1975 : Initial bifurcation process in natural channel networks, *Geology*, 3, 307-8.

Abrahams, A.D., 1976 : Evolutionary changes in link-lengths : Further evidence for stream abstraction, *Transactions of the Institute of British Geographers*, New Series 1, 225-30.

Abrahams, A D., and Campbell, R.N., 1976 : Source and tributary-source link-lengths in natural channel networks, *Bulletin of the Geological Society of America*, 87, 1016-20.

Abrahams, A.D., 1980 : Channel link density and ground slope, Annals, Association of American Geographar, 70, 80-93.

Agrawal, V.M., 1972 : Morphometric analysis of some drainage of Jabalpur Plateau, *Journal of the National Geographical Society of India*, 17, 148-68.

Anderson, H.W., 1957 : Relating sediment yield to watershed variables, *Transactions of the American Geophysical Union* 38, 921-4.

Anderson, M.G., 1973 : Measure of three-dimensional drainage basin form, *Water Resources Research*, 9, 378-83.

Aronovici, V.S., 1966 : The area-elevation ratio curve as a parameter in watershed analysis, *Journal of Soil and Water Conservation* 21, 226-8.

Asthana, V.K., 1975 : The drainage characteristics of basins around Almora and its environs, *National Geographer*, 10, 87-103.

Bassett, J.L. and Ruhe, R.V., 1973 : Fluvial geomorphology, in Karst terrain, in Morisawa, M.E. (ed), *Fluvial Geomorphology*, Binghampton, New York : University of New York Press.

Bedient, P.B., Huber, W.C., and Heaney, J.N., 1977 : Environmental model of Kissimmee river basin, *Journal, Water Resources Planning and Management Division, American Society of Civil Engineer*, 103 (WR2), 241-56.

Blair, D J. and Biss, T.H., 1967 : *The measurement of shape in geography*, Nottingham : Quantitative Bulletin No. 11, Dept. of Geography.

Boulton, A.G., 1965 : Morphometric analysis of river basin characteristics, *Water Resources Board (UK)*.

Bowden, K.L. and Wallis, J.R., 1964 : Effect of stream ordering techniques on Horton's laws of drainage composition, *Bulletin of the Geological Society of America*, 75, 767-74.

Brunsden, D., 1969 : *Dartmoor*, Geographical Association.

Brunsden, D. and Thornes, J.B., 1979 : Landscape sensitivity and change, *Transactions, Institute of British Geographers, New Series*, 4, 463-84.

Cannon, P.J., 1976 : Generation of explicit parameters for a quantitative geomorphic study of the Mill Creek drainage basin, *Oklahoma Geology Notes*, 36, 3-17.

Carson, M., 1966 : Some problems with the use of correlation techniques in morphometric studies, in Slaymarker, H.O., editor, 1966, *British Geomorphological Research Group Occasional Paper* 4.

Chayes, F., 1971 : *Ratio Correlation*, University of Chicago : Chicago.

Chorley, R.J., 1957a : Illustrating the laws of morphometry, *Geological Magazine*, 94, 140-50.

Chorley, R.J., 1957b : Climate and morphometry, *Journal of Geology*, 65, 628-38.

Chorley, R.J., 1958 : Group operator variance in morphometric work with maps, *American Journal of Science*, 256, 208-18.

Chorley, R.J., and Dale, P.F., 1972 : Cartographic problems in stream channel delineation, *Cartography*, 7, 150-62.

Christofoletti, A., 1969 : Analise morfometrica de bacias hidrograficas, *Noticia Geomorfologica* 9, 35-64.

Cunha, S.B., Machado, M.B. and Moushino de Meis, M.R., 1975 : Drainage basin morphometry on deeply weathered bedrocks, *Zeitschrift fur Geomorphologie*, 19, 125-39.

Dacey, M.F. and Krumbein, W.C., 1971 : Comments on spatial randomness in dendritic stream channel networks, *Evanston, Illinois : Northwestern University, Department of Geological Sciences, Technical Report 17, Contract no. 0014-67-A0356-0018.*

Dacey, M.F. and Krumbein, W.C., 1976 : Topological properties of disjoint channel networks within enclosed regions, *Mathematical Geology*, 8, 429-61.

Delleur, J.W. and Lee, M.T., 1973 : A rainfall-runoff model based on the watershed stream network, *Proceedings of the symposium on the design of water resources projects with inadequate data*, Vol. 1, Madrid, 75-87.

Douglas, I., 1968 : Erosion in the Sungei Gombak catchment, Selangor, Malaya, *Journal of Tropical Geography*, 26, 1-16.

Drummond, R.R., 1974 : When is a stream a stream? *Professional Geographer*, 26, 34-7.

Dunne, T., 1976 : Field studies of hillslope flow processes, in Kirkby, M.J. editor, *Hillslope Hydrology*, Chichester : Wiley.

Ebisemiju, F.S., 1979 : A reduced rank model of drainage basin morphology, *Geografisker Annaler*, 61A, 103-12.

Eyles, R.J., 1966 : Stream representation on Malayan maps, *Journal of Tropical Geography* 22, 1-9.

Faniran, A., 1969 : The index of drainage intensity : A provisional new drainage factor, *Australian Journal of Science* 31, 328-30.

Ferguson, R.I., 1977 : On determining distances through stream networks, *Water Resources Research* 13, 672-4.

Ferguson, R.I., 1978 : Drainage density-basin area relationship : Comment, *Area* 10, 350-2.

Flint, J.J., 1976 : Link slope distribution in channel networks, *Water Resources Research* 12, 645-54.

Flint, J.J., 1980 : Tributary arrangements in fluvial systems, *American Journal of Science* 280, 26-45.

Fok, Y.S., 1971 : Law of stream relief in Horton's stream morphological system, *Water Resources Research* 7, 201-3.

Gardiner, V., 1971 : A drainage density map of Dartmoor, *Reports and Transactions of the Devonshire Association for the Advancement of Science* 103, 167-80.

Gardiner, V., 1973 : The univariate distributional characteristics of some morphometric variables, *Geografisker Annaler Series* A 54A, 147-53.

Gardiner, V., 1974 : Land form and land classification in north-west Devon. *Transactions of the Devonshire Association* 106, 141-53.

Gardiner, V., 1975 : Drainage basin morphometry, *British Geomorphological Research Group Technical Bulletin* 14, Norwich : Geo-abstracts.

Gardiner, V., 1976 : Land evaluation and the numerical delimitation of natural regions, *Geographia Polonica* 34, 11-30.

Gardiner, V., 1977 : Estimated drainage density and physical regions in south-west England, *National Geographer* 12, 115-30.

Gardiner, V., 1978 : Redundancy and spatial organisation of drainage basin form indices, *Transactions of the Institute of British Geographers, New Series* 3, 416-31.

Gardiner, V., 1979 : Estimation of drainage density from topological variables, *Water Resources Research* 15, 909-17.

Gardiner, V., in press : Drainage basin morphometry, in Goudie, A. editor, *A manual of geomorphological techniques*, London : Allen and Unwin.

Gardiner, V. and Gardiner, G., 1979 : *Analysis of Frequency Distributions*, Norwich : Geo-abstracts.

Gardiner, V. and Park, C.C., 1978 : Drainage basin morphometry : Review and assessment, *Progress in Physical Geography* 2, 1-35.

Gardiner, V. Gregory, K.J. and Walling, D.E., 1977 : Further notes on the drainage density-basin area relationship, *Area* 9, 117-21.

Gardiner, V., Gregory, K.J. and Walling, D.C., 1978 : Further notes on the drainage density-basin area relationships : Comment reply, *Area* 10, 354-5.

Ghose, B. and Pandey, S., 1963 : Quantitative geomorphology of drainage basins, *Journal of the Indian Society of Soil Science* 11, 259-74.

Ghose, B. Pandey, S. and Lal, G., 1967 : Quantitative geomorphology of the drainage basins in the central Luni basin in western Rajasthan, *Zeitschrift fur Geomorphologie I*, 146-60.

Ghosh, A.K. and Scheidegger, A.D., 1970 : Dependence of stream link lengths and drainage areas on stream orders, *Water Resources Research* 6, 336-40.

Goudie, A., 1969 : Statistical laws and dune ridges in southern Africa, *Geographical Journal* 135, 404-6.

Graf, W.L., 1975 : A cumulative stream-ordering system, *Geographical Analysis* 7, 335-40.

Gregory, K.J., 1966 : Dry valleys and the composition of the drainage net, *Journal of Hydrology*, 4, 327-40.

Gregory, K.J., 1976a : Drainage networks and climate, in Derbyshire, E., editor, *Geomorphology and climate*, London : Wiley.

Gregory, K.J., 1976b : Drainage density changes in south-west England, in Gregory, K.J. and Ravenhill, W.L.D., editors, *Exeter Essays in Geography*, Exeter.

Gregory, K.J., 1976c : Pragmatic river channel typology, *International Geography*, 76 I, 158-60.

Gregory, K.J., 1977 : Stream network volume—an index of channel morphometry, *Bulletin of the Geological Society of America* 88, 1075-80.

Gregory, K.J., 1978 : Fluvial processes in British basins, in Embleton, C., Brunsden, D. and Jones, D.K.C. editors, *Geomorphology : Present Problems and Future Prospects*, Oxford.

Gregory K.J., 1979a : Drainage network composition, *Acta Universitatis Ouluensis*, series A, 82, 19-28.

Gregory, K.J. and Gardiner, V., 1975 : Drainage density and climate, *Zeitschrift fur Geomorphologie* 19, 287-98.

Gregory, K.J. and Gradiner, V., 1979 : Comment on "Drainage density and streamflow : A closer look", by S.L. Dingman, *Water Resources Research* 15, 1662-4.

Greogory, K.J. and Walling, D.E., 1968 : The variation of drainage density within a catchment, *Bulletin of the International Association of Scientific Hydrology* 13, 61-8.

Gregory, K.J. and Walling, D.E., 1973 : *Drainage basin form and process*, London : Edward Arnold.

Gupta, A. and Chakraborty, S.C., 1967 : Quantitative analysis of landforms—a case study on the Dulong Basin, *Proceedings of Seminar on Geomorphological Studies in India*, 190-9.

Gupta, S.N., Bhattacharya, A. P. and Jindal, S.R., 1969 : Statistical correlations of Himalayan and Bundelkhand basin characteristics with flood flows, *International Association of Scientific Hydrology, Publication* 85, 874-80.

Gurnell, A.M., 1978 : The dynamics of a drainage network, *Nordic Hydrology* 9, 293-306.

Gustafson, E.C., 1973 : Quantitative investigation of the morphology of drainage basins using orthophotography, *Munchener Geographische Abhandlungen* 11.

Hack, J.T., 1957 : Studies of longitudinal stream profiles in Virginia and Maryland, *United States Geological Survey Professional Paper* 294-B 45-97.

Haggett, P. and Chorley, R.J., 1969 : *Network Analysis in Geography*. London : Edward Arnold.

Harlin, J.M., 1978 : Statistical moments of the hypsometric curve and its density function, *Mathematical Geology* 10, 59-72.

Haynes, V.M., 1972 : The relationship between the drainage areas and sizes of the outlet troughs of the Sukkertoppen ice cap, West Greenland, *Geografisker Annaler* 54A, 66-75.

Horton, R.E., 1926 : Flood flows characteristics : Discussion, *Transactions of the American Society of Civil Engineers* 89, 1081-6.

Horton, R.E., 1932 : Drainage basin characteristics, *Transactions of the American Geophysical Union* 13, 350-61

Horton, R.E. : Erosional development of streams and their drainage basins : Hydrophysical approach to quantitative morphology, *Bulletin of the Geological Society of America* 56, 275-370.

James, W.R. and Krumbein, W.C., 1969 : Frequency distributions of stream link-lengths, *Journal of Geology* 77, 544-65.

Jarvis, R.S., 1976a : Link-length organisation and network scale dependencies in the network diameter model, *Water Resources Research* 12, 1215-25.

Jarvis, R.S., 1976b : Classification of nested tributary basins in analysis of drainage basin shape, *Water Resources Research* 12, 1151-64.

Jarvis, R.S., 1977a : Drainage network analysis, *Progress in Physical Geography* 1, 271-95.

Jarvis, R.S., 1977b : Spatial variation in ordered tributary basins—a proposed classification, *Geografisker Annaler* 59A, 241-51.

Jarvis, R.S., in press : Specific geomorphometry, in Goudie, A., editor, *A manual of geomorphological techniques*, London : Allen and Unwin.

Jarvis, R.S. and Werrity, A., 1975 : Some comments on testing random topology stream network models, *Water Resources Research* 11, 309-18.

Jennings, J.N., 1973 : 'Any milleniums today lady?' The geomorphic bandwagon parade, *Australian Geographical Studies* 11, 115-33.

Kirchner, W.B., 1975 : An examination of the relationships between drainage basin morphology and the export of phosphorus, *Limnology and Oceanography*, 20, 267-79.

King, R.B., 1970 : A parametric approach to land system classification, *Geoderma* 4, 37-46.

King, R.B., 1975 : Geomorphic and soil correlation analysis of land systems in Northern and Laupula Provinces of Zambia, *Transactions of the Institute of British Geographers* 64, 67-76.

Kirkby, M.J., 1976 : Testes of the random network model and its application to basin hydrology, *Earth Surface Processes* 1, 197-212.

Kumar, A. and Pandey, R.N., 1975 : Morphometric properties of the polygenetic drainage basin : A case study of the Burha Nadi, *Geographical Review of India* 37, 329-38.

Langbein, W.B., 1947 : Topographic characteristics of drainage basins, *United States Geological Survey Water Supply Paper* 968-C, 125-57.

Leopold, L.B., 1971 : Trees and streams : The efficienacy of branching patterns, *Journal of Theoretical Biology* 31, 339-54.

Leopold, L.B. and Langbein, W.B., 1962 : The concept of entropy in landscape evolution, *United States Geological Survey Professional Paper* 500-A. (20 pp.)

Maner, S. B., 1958 : Factors affecting sediment delivery rates in the Red Hills physiographic area, *Transactions of the American Geophysical Union* 39, 669-75.

Mark, D.M., 1974 : Line intersection method for estimating drainage density, *Geology* 2, 235-6.

Mark, D.M., 1979 : Topology of ridge patterns : Randomness and constraints, *Bulletin of the Geological Society of America* Part 1, 90, 164-72.

Mark D.M. and Church, M., 1977 : On the misuse of regression in earth science, *Mathematical Geology* 9, 63 -75.

Mather, P. and Doornkamp, J.C., 1970 : Multivariate analysis in geography, with particular reference to drainage basin morphometry, *Transactions of the Institute of British Geographers* 51, 163-87.

McArthur, D.S. and Ehrlich, R., 1977 : An efficiency evaluation of four drainage basin shape ratios, *Professional Geographer* 29, 290-5.

Melton, M.A., 1958a : Geometric properties of mature drainage systems and their representation in an E4 phase space, *Journal of Geology* 66, 25-54.

Melton, M.A., 1958b : Correlation structure of morphometric properties of drainage systems and their controlling agents, *Journal of Geology* 66, 442-60.

Milling, M.E. and Tuttle, S.D., 1964 : Morphometric study of two drainage basins near Iowa City, Iowa. *Transactions of the Iowa Academy of Science* 71, 304-19.

Milton, L. E., 1966 : The Geomorphic irrelevance of some drainage net laws, *Australian Geographical Studies* 4, 89-95.

Mithal, R.S., Parkash, B. and Bajpai, I.P., 1974 : Drainage basin morphometric study of a part of the Garhwal Himalayas, *Himalyan Geology* 4, 195-215.

Mock, S.J., 1971 : A classification of channel links in streams networks, *Water Resources Research* 7, 1558-66.

Morgan, R.P.C., 1971 : A morphometric study of some valley systems on the English chalklands, *Transactions of the Institute of British Geographers* 54, 33-44.

Morgan, R.P.C., 1976 : The role of climate in the denudation system : A case study from West Malaysia, in Derbyshire, E., editor, *Geomorphologyl and Climate*, London : Wiley, 319-44.

Morisawa, M.E., 1957 : Accuracy of determination of stream lengths from topographic maps, *Transactions of the American Geophysical Union* 38, 86-8.

N.E.R.C., 1975 : Natural Environment Research Council Flood Studies Report, 5 volumes, London.

Neueschwander, C., 1944 : Morphometrische Begriffe, *Arbeiten aus dem Geographische Institute d' Universitat Zurich, Serie* A, 2.

Oeppen, B.J and Ongley, E.D , 1975 : Spatial point processes applied to the distribution of river junctions, *Geographical Analysis*, 7, 153-71.

Onesti, L.J. and Miller, T.K., 1974 : Patterns of variation in a fluvial system, *Water Resources Research* 10, 1178-86.

Onesti, L.J. and Miller, T.K., 1978 : Topological classifications of drainage networks : An evaluation, *Water Resources Research* 14, 144-8.

Ouma, J.P.B.M., 1970 : Fluvial differentiation of the basin of Lake Victoria, *Journal of Hydrology* (*New Zealand*) 9, 219-51.

Ovenden, J.C. and Gregory, K.J., 1980 : The permanence of stream networks in Britain, *Earth Surface Processes* 5, 47-60.

Padmaja, G., 1975 : Some aspects of quantitative drainage characteristics of the Dhund basin, *Geographical Review of India* 37, 158-64.

Park, C.C., 1976 : Stream channel morphology in Mid-Devon, *Reports and Transactions of the Devonshire Association for the Advancement of Science* 107, 25-41.

Patton, P.C. and Baker, V.R., 1976 : Morphometry and floods in small drainage basins subject to diverse hydrogeomorphic controls, *Water Resources Research* 12, 941-52.

Pethick, J.S., 1974 : A note on the drainage density—basin area relationship, *Area* 7, 217-22.

Pike R.J. and Wilson, S.E., 1971 : Elevation-relief ratio, hypsometric integral and geomorphic area-altitude analysis, *Bulletin of the Geological Society of America* 82, 1079-84.

Prasad, C. and Verma, V.K., 1975 : Studies in basin morphometry and slope development in the western Doon valley, Garhwal Himalayas, *Himalayan Geology* 5, 425-34.

Proctor, C.V. Jr., 1969 : The North Bosque Watershed : Inventory of a drainage basin, Waco : Baylor University, *Baylor Geological Studies, Bulletin* 16.

Richards, K.S., 1977 : Slope form and basal stream relationships : Some further comments, *Earth Surface Processes* 2, 87-92.

Richards, K.S., 1979 : Prediction of drainage density from surrogate measures, *Water Resources Research* 15, 435-42.

Robinson, G., 1967 : The geomorphic irrelevance of some drainage net laws—comment, *Australian Geographical Studies* 5, 81-2.

Ruhe, R.V., 1952 : Topographic discontinuities of the Des Moines lobe, *American Journal of Science* 250, 46-56.

Schumm, S.A., 1956 : The evolution of drainage systems and slopes in badlands at Perth Amboy, New Jersey, *Bulletin of the Geological Society of America* 67, 597-646.

Schumm, S.A., 1977 : *The Fluvial System*, London : Wiley.

Schumm, S.A. and Lichty, R., 1965 : Time, space and causality in geomorphology, *American Journal of Science* 263, 110-19.

Selby, M.J., 1968 : Morphometry of drainage basins in areas of pumice lithology, *Proceedings of the fifth New Zealand Geographical Conference, New Zealand Geographical Society* 169-74.

Sharma, H.S. and Padmaja, G., 1977 : Quantitative geomorphic characteristics of Streams of the Morel Basin Rajasthan, *Geographical Review India*, Vol. 39 pp. 353-366.

Shreve, R.L., 1966 : Statistical law of stream numbers, *Journal of Geology* 74, 17-37.

Shreve, R.L., 1965 : Infinite topologically random channel networks, *Journal of Geology* 75, 178-86.

Shreve, R.L., 1974 : Variation of mainstream length with basin area in river networks, *Water Resources Research* 10, 1167-77.

Shreve, R.L., 1975 : The probabilistic approach to drainage basin geomorphology, *Geology* 3, 527-9.

Shukla, S. and Verma, V.K., 1974 : Geomorphostatistical investigations in a part of Doon valley, Garhwal Himalayas, *Himalayan Geology* 4, 229-39.

Singh, L., 1971 : Ruggedness number as a parameter in terrain regionalization: A case study of Hoshiarpur Kangra Tract, *Indian Geographical Journal* 46, 17-21.

Singh, R.P. and Kumar, A., 1969 : Geomorphological evolution of stream order of the Topa and Silphi basins in Ranchi, *Journal, National Geographical Society of India* 15, 38-44.

Singh, S., 1976 : Morphometric determinants of the stage of the cycle of erosion, *Geographical Observer* 12, 30.

Singh, S., 1979 : Shape analysis of open and closed links of small drainage basins of Ranchi Plateau, India, *Geographical Observer* 15, 1-6.

Singh, S. and Ghose, B., 1973 : Interrelationships between quantitative geomorphic characteristics of the drainage basins in sub-humid to humid environment of Rajasthan, Annals of Arid Zone 12; 82-9.

Singh, S. and Srivastava, R., 1974 : A morphometric study of the tributary basins of upper reaches of the Belan river, *National Geographer* 9, 31-44.

Singh, S. Ullah, W., Ghose, B. and Kaith, D.S., 1969 : Interpretation of aerial photographs of quantitative geomorphic characteristics of the drainage basins in sub-humid to humid environment, *Indian Journal of Geography* 4, 1-11.

Smart, J.S., 1970 : Use of topologic information in processing data for channel networks, *Water Resources Research* 6, 932-6.

Smart, J.S., 1972 : Quantitative characterization of channel net structure, *Water Resources Research* 8, 1487-96.

Smart J.S , 1973a : Environmental constraints on the substitution of space for time in the study of natural channel networks : Discussion, *Bulletin of the Geological Society of America* 84, 351-2.

Smart, J.S., 1973b : The random model in fluvial geomorphology, in Morisawa, M.E., editor, *Fluvial geomorphology*, Binghampton, New York.

Smart, J.S., 1976 : Joint distribution functions for link-lengths and drainage areas, in Merriam, D.F., editor, *Random Processes in Geology*, Springer-Verlag, New York.

Smart, J.S., 1978 : The analysis of drainage network composition, *Earth Surface Processes* 3, 129-70.

Smart, J.S. and Surkan, A.J., 1967 : The relation between mainstream length and area in drainage basins, *Water Resources Research* 3, 963-74

Smart, J.S. and Werner, C., 1976 : Applications of the random model of drainage basin composition, *Earth Surface Processes* 1, 219-33.

Smart, J.S. and Wallis, J.R., 1971 : Cis and Trans links in natural channel networks, *Water Resources Research* 7, 1346-8.

Smith, K.G., 1950 : Standards for grading texture of erosional topography, *American Journal of Science* 248, 655-68.

Strahler, A.N., 1952 : Hypsometric (area-altitude) analysis of erosional topography, *Bulletin of the Geological Society of America* 63, 1117-42.

Strhaler, A.N., 1958 : Dimensional analysis applied to fluvially eroded landforms, *Bulletin of the Geological Society of America* 69, 279-300.

Strhaler, A.N., 1964 : Quantitative geomorphology of drainage basins and channel networks, in Chow, V. T., editor, *Handbook of applied hydrology*, 4, 39-4, 76.

Surkan A.J., 1974 : Simulation of storm velocity effects on flow from distributed channel networks, *Water Resources Research* 10, 1149-60.

Tandon, S.K., 1971 : A study in the stratigraphic frequency distributions of various third order basin parameters in a part of the Kumaun Himalayas, *Journal, National Geographical Society of India.* 17, 127-33.

Tandon, S.K., 1974 : Litho-control of some geomorphic properties : An illustration from the Kumaun Himalayas, India, *Zeitschrift fur Geomorphologie* 18, 460-71.

Thomas, M.F., 1969 : Geomorphology and land classification in tropical Africa, in Thomas, M.F. and Whittington, G.W., editors, *Environment and land use in Africa*, London : Methuen.

Thomas, M.F., 1976 : Purpose, scale and method in land resources surveys, *Geographia Polonica* 34, 207-24.

Vaidyanadhan, R., 1977 : Recent advances in geomorphic studies of Peninsular India : A review, *Indian Journal of Earth Sciences* (S. Ray volume), 13-35.

Verma, V.K., 1972 : Drainage and slope characteristics of lower Gondwana terrain of central India, *Noticia Geomorfologica* 12, 39-50.

Verstappen, H. Th., 1977 : *Remote sensing in Geomorphology*, Amsterdom : Elsevier.

Werner, C., 1972 : Channel and ridge networks in drainage basins, *Proceedings of the Association of American Geographers* 4, 109-14.

Werner, C., 1973 : The boundaries of drainage basins for topologically random channel networks, *Proceedings of the Association of American Geographers* 5, 287-90.

Werner, C. and Smart, J.S., 1973 : Some new methods of topologic class'fication of channel networks, *Geographical Analysis* 5, 271-95.

Werrity, A., 1972 : Accuracy of stream link-lengths derived from maps, *Water Resources Research* 8, 1255-64.

Wilcock, D.N., 1975 : Relations between planimetric and hypsometric variables in third and fourth order basins, *Bulletin of the Geological Society of America* 86, 47-50.

Williams, P.W., 1972 : The analysis of spatial characteristics of Karst terrains, in Chorley, R.J., editor, *Spatial analysis in geomorphology*, London : Methuen.

Woldenberg, M.J., 1967 : Geography and properties of surfaces, *Harvard Papers in Theoretical Geography* 1, 95-189.

Woldenberg, M.J. and Berry, B.J.L., 1967 : Rivers and central places : Analogous systems, *Journal of Regional Science* 7, 129-39.

Wong, S.T., 1975 : A deductive approach for the prediction of mean annual flood in a humid region, in *The hydrological charccteristics of river basins and the effects on these characteristics of better water management*, Pub. 117, International Association for Scientific Hydrology, Tokyo.

Woodyer, K.D. and Brookfield, M., 1966 : The land systems and its stream net, CSIRO Division of Land Resources, Technical Memorandum 66/5.

SEVEN

H. S. Sharma and *G. Padmaja*

QUANTITATIVE FLUVIAL CHARACTERISTICS OF STREAMS OF THE MEJ BASIN (RAJASTHAN)

In the recent years considerable conceptual change has taken place in geomorphology. Sophisticated quantitative techniques have been evolved to analyse the dynamic nature of earth's form. Various morphometric techniques through statistics are becoming more and more useful to a geomorphologist to analyse the land as an open system. The studies in this aspect of research have been initiated by Horton (1945), an eminent hydrologist and Strahler (1950, 52, 54, 56, 58) and are widely publicised by other geomorphologists like Kesseli (1946), Russell (1949), Miller (1953), Schumm (1956), Melton (1958), Smith (1958), Morisawa (1962), Robinson (1963), Leopold, Wolmen and Miller (1964), Clarke (1966), Chorley (1966) and number of others.

Though the study of quantitative drainage analysis has advanced with rapidity in American and European countries, the follow up of this branch in India is only of recent origin and the progress has been comparatively slower. Names of Ghose, B. Pandey and Singh (1967), Sharma, V. (1968), Verma and Tandon (1971), Tandon (1974) and Pal, S.K. (1973) are a few who have publicized and tested Horton's and Strahler's models in different parts of this country.

The present work is an attempt to study the numerical indices of the Mej Basin and to try out the different laws propounded by Horton and Strahler. The Mej river is a main left bank tributary of the Chambal. It forms an oval-shaped basin over an area of 5,500 square kilometres. The Basin is a transitional zone between the north-western lobe of the Great Vindhyan Basin and south-eastern fringe of the Aravallis. Geographically, it extends from 25°′ to 25° 45′ north latitude and 75°30′ to 76°15′ east longitudes. (*Fig.7.1*)

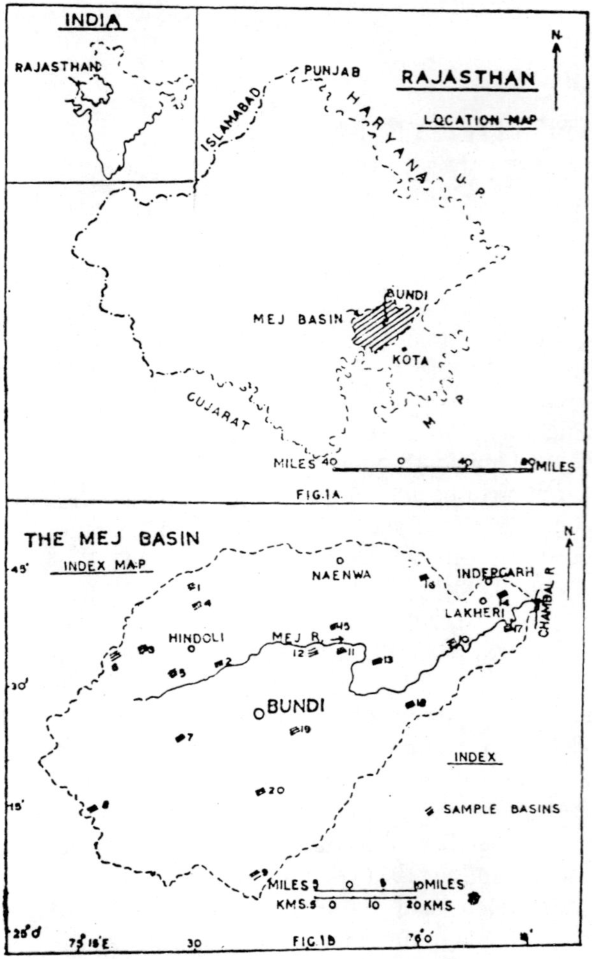

FIG. 7.1 : *Rajasthan location map and the Mej Basin*

Climatically, it falls under semi-arid zone of India. According to the Thornthwaites (1948) moisture index, the Basin has a megathermal type of climate.

Broadly, the Mej Basin can be divided into three topographic units—Hills, Plateaus and Plains. The hilly traverse from south-west to north-east and like a backbone divide the basin into two equal parts. The northern area to the north of these hills is a plateau having an average elevation 330 metres. South-western portion of the Basin is an extension of the Kota plateau. On the other hand, the south-eastern part of the basin is a plain overlain by recent and subrecent sediments deposited by the Mej and its tributaries. The plain is generally below 300 metres, the average height being 210 metres. This area is characterized by dissected ravines and gullies.

Material and methods

The present study is based on field study. Apart from the general observation in the field, various topographical sheets were consulted for detailed morphometric analysis. For the quantitative analysis of fluvially eroded topography twenty representative basins from different lithological conditions were selected, as is evident from the Table 7.1. Log and semilogpapers have been used to depict different laws of drainage basins propounded by Horton and Strahler. Length of streams and area of basins were calculated by map measurer and planimeter respectively.

TABLE 7.1 : **Lithological conditions**

Lithological group	1 Gawaliors	2 Upper Bhander Vindhyan Sandstone	3 Vindhyan Sand-stone, Shales, Limestones alluvium	4 Gwaliors Kaimur Sandstone, Alluvium	5 Alluvium
Basin Number	1,2,3,4, 5, 6	7,8,9,	10,11,12, 13,14	15,	16,17,18, 19,20

Stream number

Horton evolved a model of stream ordering and suggested that the number of channels of each order of a given basin form an inverse geometric series with the order. The Horton's model of stream number can be expressed by the following equation :

$$N_u = R_b\,(S - U)$$

where

R_b denotes bifurcation ratio,

S represents the higher order of the basin.

This has been termed as first law of drainage composition and has been widely tested in different parts of the world. In this Basin, the streams were ordered according to Strahler's method, where all the fingertip tributaries were named as order 1 and in turn 2 first order streams join to make a stream of second order and so on.

The data of twenty sample basins of the Mej Basin exhibit the orderly arrangement of the number of streams of different orders. It is evident from the plots on a semilog paper (Fig. 7.2) that the number of streams decrease with an increase in order, thereby the exponential function is negative. The points generally fall close to a straight line. The common negative exponential function of different basins is expressed by the following equation :

$$Y = 43.24 - 11.6X$$

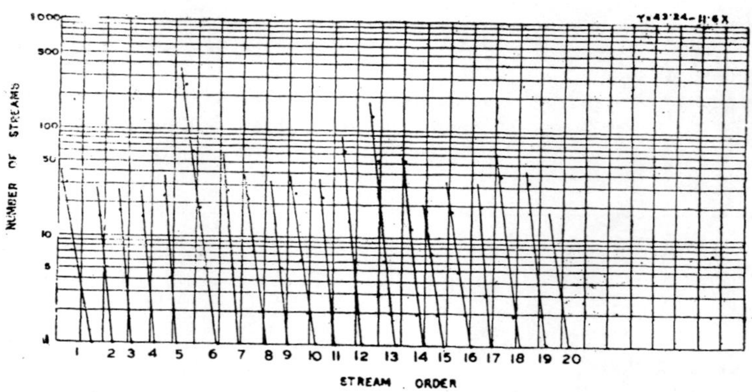

FIG. 7.2 : *The number of streams with order*

When the average of stream numbers in all the lithological groups is taken, the higher number of streams fall in the category of the Gwaliors (21.09) and Vindhyan sandstone and alluvium (20.3). Minimum number of streams (3.3) fall in group 4 (Gwaliors and Kaimur sandstone). The average value in alluvium comes to 8.4 which is because of the less number of streams on an average along with higher number of sample basins in this group.

Bifurcation ratio (R_b)

Horton suggested that the slope of the fitted lines represent the mean bifurcation ratio of the series. He considered the bifurcation ratio as an index of relief and dissection. Number of workers pointed out that the bifurcation ratio on mature surfaces tend to have values between 3 and 5, with an usual value around 4. Bifurcation ratio is the ratio of the total number of streams of one order to that of the next higher order. Because of chance irregularities, bifurcation ratio between successive pairs of orders differs, within the same basin even if a general observance of a geometric series exists (Schumm, 1956). To arrive at a more representative bifurcation number, Strahler (1953) used a weighted mean bifurcation ratio. The calculated weighted mean bifurcation ratio is tabulated in Table 7.2.

TABLE 7.2 : **Weighted mean bifurcation ratio in different lithological groups**

Gwaliors	Upper Bhander Vindhyan Sandstone	Vindhyan Sandstone, Limestone, Alluvium	Gwalior Sandstone, Alluvium	Kaimur Alluvium
4.004	—	—	—	—
3.46	5.17	3.79	—	3.22
4.20	3.14	5.48	3.125	6.01
3.46	3.20	4.47	—	4.83
3.54	—	3.97	—	3.96
3.96	—	4.47	—	2.52

The Table 7.2 indicates that the value of bifurcation ratio varies from a least value of 2.52 to 6.01, both falling in the alluvium group, otherwise the values range around 4 as has been suggested earlier. The exceptionally higher value of weighted mean bifurcation ratio in the alluvium group is because of more number of streams in the form of gullies and also because of the fact that the variation in number of streams from one order to the other is more.

The highest value (3.96) of R_b falls in Vindhyan sandstone, limestone and alluvium and the lowest value (2.75) is in the group of the Gwaliors, Kaimur sandstone and alluvium. The higher values for group three is merely because of more number of streams and also being a higher ordered basin. Strahler (1952) has stated that the bifurcation ratio shows a smaller range of variation from region to region except where powerful geological character-predominates. Miller (1953), further found out that lithological variations do not cause differences in bifurcation ratio. The average mean weighted bifurcation ratio ranges from 3.125 in lithological group four to 4.43 in group three supporting the above views of Strahler and Miller, except basin 13 (8.50) where powerful geological control exists.

The comparative study of mean bifurcation ratio and weighted mean bifurcation ratio indicates that the values of weighted mean bifurcation ratio range slightly higher than the mean bifurcation ratio, in all the lithological groups.

Stream length

The second law of Horton concerns stream lengths. He found out that the mean length of stream segments of a given order increases exponentially with increasing order number.

In the present study when the mean stream length is plotted against order (Fig. 7.3) it shows a positive exponential function. In this case, though the points fall away from the fitted lines, but the regression lines give an appearance of positive exponential function, which is expressed by the following equation :

$$Y = .08 + .5X$$

The plots of total stream length against order (Fig. 7.4) give a contrast result. These lines show a negative exponential function, where the total stream length decreases with an increase in order. The common equation for this relation is :

$$Y = 14.38 - 3.06X$$

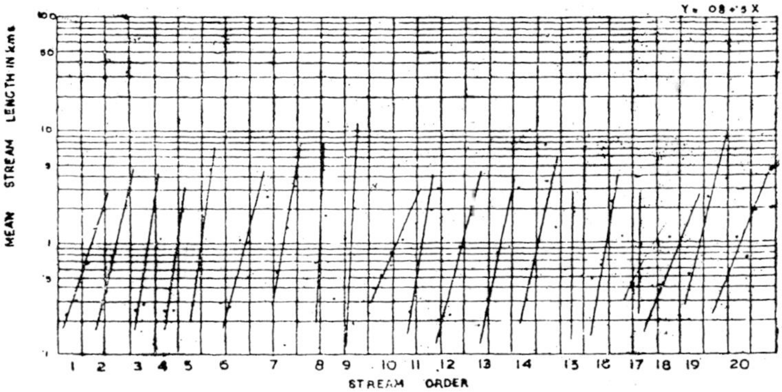

FIG. 7.3 : *The mean stream length is plotted against order*

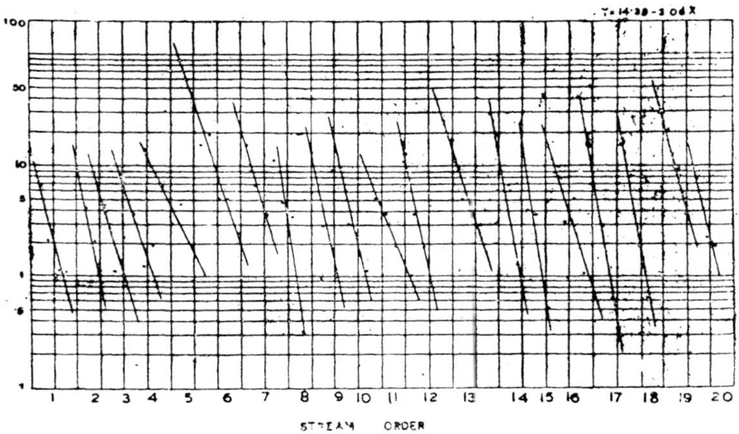

FIG. 7.4 : *Plot of total streams length against order*

The cumulative mean and total stream lengths have also been plotted against order. (Fig. 7.5 and Fig. 7.6). Both these graphs show a common positive exponential function. The equation of the cumulative mean stream length against order is :

$$Y = -0.36 + 1.16X$$

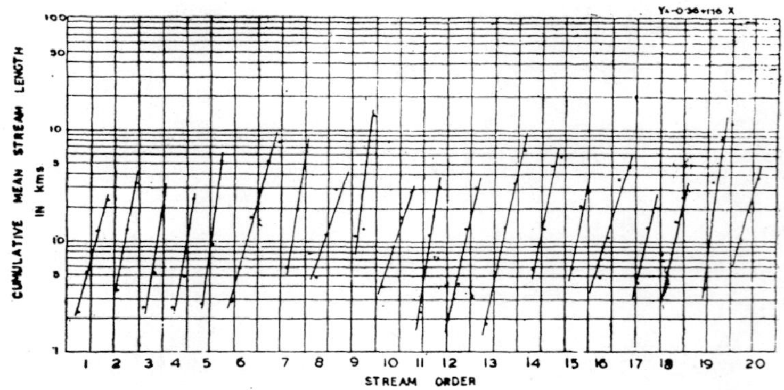

FIG. 7.5 : *Cumulative mean and total lengths plotted against order*

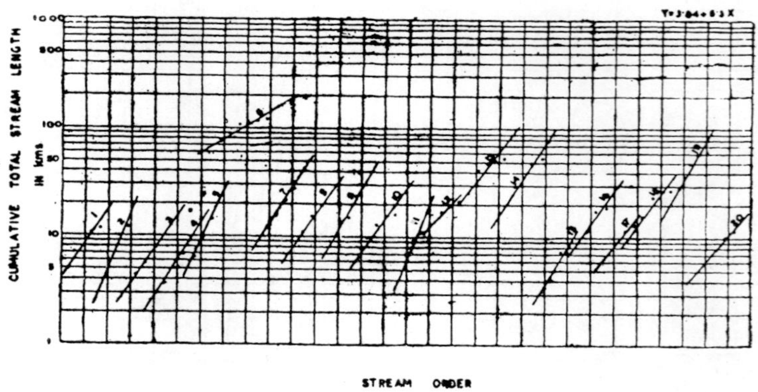

FIG. 7.6 : *Cumulative total stream length*

And for cumulative total stream length against order is (Fig. 7.6)

$$Y = 3.84 + 6.3X$$

Channel length ratio

Similar to bifurcation ratio, the channel length ratio is also a ratio of the length of a stream of an order to the stream length of its previous order. Use of total stream length and mean stream length for all the twenty basins yields different results. Channel ratio from the total stream length has the highest value (9.60) in basin 9 of Vindhyan Sandstone group between second and third orders. This is because of the fact that the length of the third order increases astonishingly as compared to a meagre value of second order streams. Otherwise, the values fall around a value of .5 in the rest of the basins. The average table also shows a higher value of 1.71 in the same group. Least value of .49 falls under the category of alluvium. The general values for all the basins also range around .4.

The Table 7.3 gives a picture of the different values of stream length ratios in various lithological groups.

TABLE 7.3 : **Stream length ratios from different lithologies**

Group 1	Group 2	Group 3	Group 4	Group 5
a. From mean stream length				
1.9	3.06	1.4	1.79	1.9
3.16	1.50	2.6	—	1.4
6.8	42.4	2.4	—	1.3
6.2	—	1.1	—	1.4
5.4	—	2.3	—	1.8
1.69	—	—	—	—
b. From total stream length				
.52	.50	.41	.63	.71
.66	.47	.64	—	.29
.92	3.7	.45	—	.42
.78	—	.58	—	.57
.91	—	.41	—	.45
.49	—	—	—	—

Channel length ratio from mean stream length has a higher value in comparison to channel length ratio from total stream length. Basin 9 again has the highest value of 48.08 between second and third orders due to similar reason mentioned above. The same basin has also the lowest value of .18 between first and second orders. This is also due to the fact that the length of first order streams is more than the second order streams. The average values also suggest the same fact that the highest value of 8.28 belongs to second group and a lowest value of 1.625 lies in group five. The weighted mean channel length ratio also has similar results of 42.4 in basin 9 and a lowest value of 1.1 in basin 13. The averages too fall in the same category of 15.65 in group two and 1.56 in group five.

The values of weighted channel length ratios from total stream length has a higher value (3.7) in basin 9 and drops down to a least value of .29 in basin 17. The average weighted channel length ratio for the same parameter also speaks of higher values (1.56) in group two and a least value of 4.88 in alluvium group. In general, it can be concluded that the stream length ratios from total and mean stream lengths are highest in group one and two and have a least value in the last group.

Bifurcation ratio and channel length ratio

The values of stream length ratios from mean and total stream lengths have been plotted against the values of bifurcation ratio. Both the plots give a inverse relation between their values. When the channel length ratio from mean stream length is plotted against bifurcation ratio (Fig. 7.7), it gives a appearance of increasing trend with increasing values of bifurcation ratios, similar to Horton's first law of stream numbers emphasizing the fact that the mean length of stream increase with increasing order. The positive exponential function of the relation is expressed in the following equation :

$$Y = .14 + .79X$$

A reverse trend is noticed when the channel length ratio from total stream length is plotted against bifurcation ratio (Fig. 7.8) confirming the fact that the total stream length of a stream decreases with an increase in order. The equation of the negative exponential function is quoted below :

$$Y = .37 - .105X$$

Basin area

The next law of drainage basins is the law of basin area. In the present study the basin areas have been calculated according to Strahler's method (1971).

Individual basin areas from total stream lengths have been plotted against order (Fig. 7.9), on a semilog paper, taking

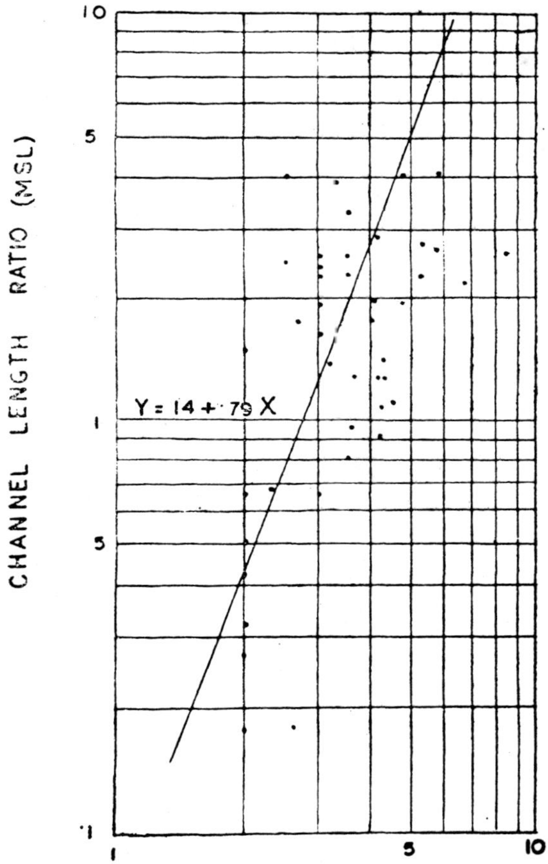

FIG. 7.7 : *Bifurcation ratio and channel length ratio (MSL)*

basin order on abscissa and basin area on ordinate. These points show an inverse relation expressed by the following equation :

$$Y = 16.00 - 2.5X$$

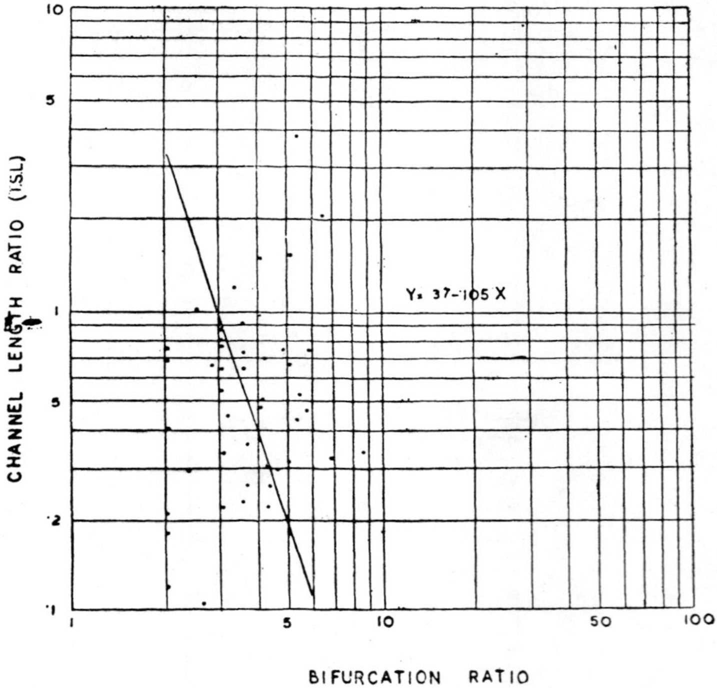

FIG. 7.8 : *Total stream length have been plotted against order ratio*

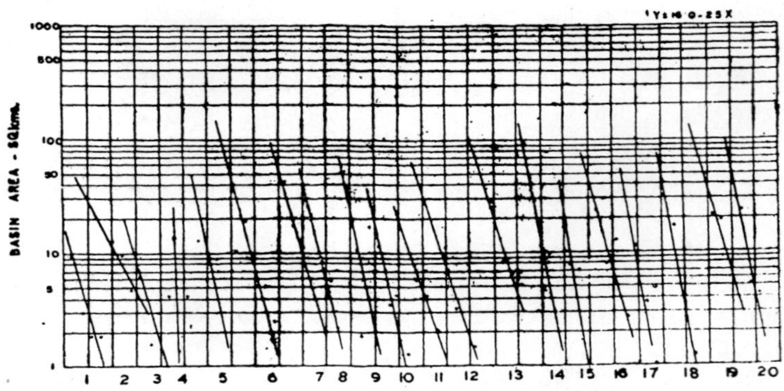

FIG. 7.9 : *Basin area (TSL)*

This decrease in area with increasing order is mainly due to the fact that the more number of finger tip streams along with inter-basin areas occupy larger areas than the main channel.

Apart from the above mentioned relation, cumulative total basin area is also plotted against order (Fig. 7.10). This relation is reverse to the former one, in which the basin area increases with an increase in order in accordance with Strahler, who is of opinion that the area of a higher order stream includes the total area covered by all the streams of lower order. The positive exponential function for the plots is given below:

$$Y = .1 + 15.3X$$

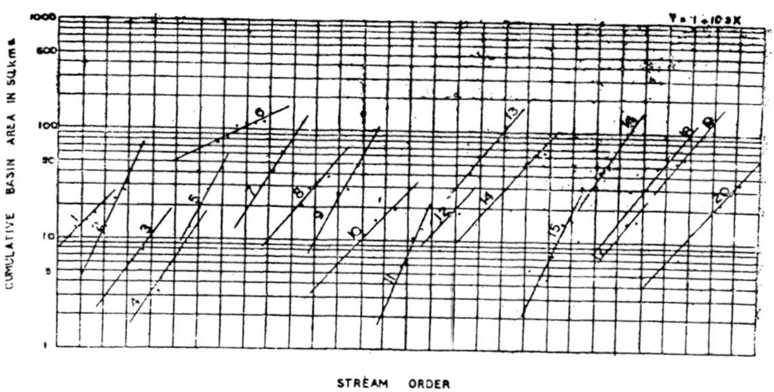

FIG. 7.10 : *Cumulative basin area plotted against order*

Cumulative mean basin area has also been related with order, the plots of which also speak of the positive exponential function of the curve (Fig. 7.11). The equation is as follows :

$$X = -1.7 + 9.42X$$

The relation once again confirms the law that the mean basin area increases with increase in order.

Area ratio

Similar to bifurcation ratio and channel length ratio, area ratio from total as well as mean stream lengths have been calculated in this study. It is found that the area ratio from mean stream length has higher values than that calculated from the total stream length, thereby depicting a similar behaviour to that of

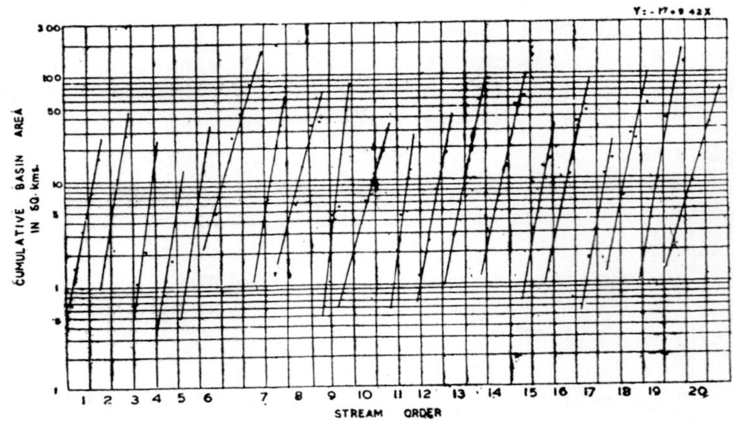

FIG. 7.11 : *Cumulative basin area plotted against order*

channel length ratio. Area ratios from total and mean stream lengths are higher in basin 9 both ranging to values of 9.2 and 44.4 kilometres respectively. But the values of area ratio from mean stream length obtain higher values in the Gwalior group (2.86 kilometres). Though the general value lies around 2, group four overtakes the value of group second by .13.

TABLE 7.4 : **Area ratios in different types of lithologies**

Group 1	Group 2	Group 3	Group 4	Group 5
a. From mean stream length				
1.86	2.8	1.4	1.66	1.9
2.40	1.3	3.3	—	.445
4.55	22.30	3.5	—	1.44
4.20	—	2.0	—	1.83
4.35	—	2.15	—	1.66
1.85	—	—	—	—
b. From total stream length				
.66	.53	.51	.51	.69
.83	.42	.61	—	.27
1.08	4.63	.55	—	.36
.86	—	.62	—	.56
1.18	—	.43	—	.49
.76	—	—	—	—

Table 7.4 indicates that the average values from mean stream length have extraordinarily high values for basin 9 (22.30) and also a value of 4.63 for basin area from total stream length and again a least value of .445 and .27 for basin 17 of alluvium group. On the whole the area ratios and stream length ratios have an identical behaviour in all the lithological groups.

Basin area and length

In all the above analysis various parameters were plotted against order but in figure number 12 and 13 cumulative basin area has been plotted against cumulative total and mean stream lengths. In figure 7.12 cumulative total stream length is plotted against cumulative total basin area. In this relation regression

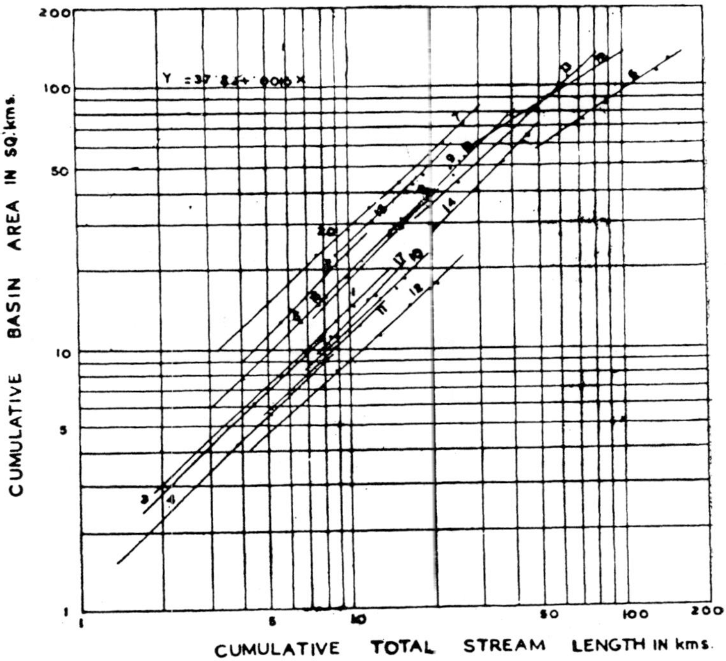

FIG. 7.12 : *Cumulative total stream length is plotted against cumulative total basin area*

lines for all the sample basins were drawn and a common regression line has also been fitted (Fig. 7.13). All these regression lines appear more or less parallel except the line of basin

19 and in general show the positive exponential function of the relation. The value of common regression line is :

$$Y = 37.83 + .0016X$$

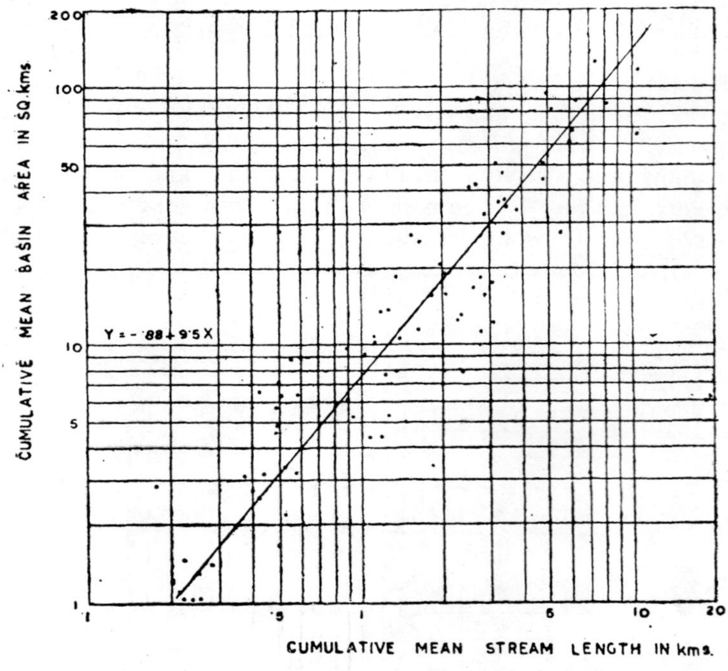

FIG. 7.13 : *Cumulative mean basin area plotted against order*

In the same manner in figure number 15 cumulative mean basin area has been plotted against cumulative mean stream length. It also shows the similar function. The value of commonly fitted regression line is :

$$Y = -.88 + 9.5X$$

Thus, it can be suggested that the both stream length and basin area show a constant ratio increase with order.

Drainage density

Drainage density is the length of a stream per square unit area, (Fig. 7.14) obtained by dividing the whole length of the streams by the area of the basin. The analysis indicates that the drainage density has its highest value in basin 6 (.97) which is under first group. Apart from this, basins 12 and 14 of group three also have a value of 1.13 and .71. A least value of .34 is found in basin 20. The high density in the Gwaliors can be referred to the fact that the Gwalior shales give a high proportion of streams. The Vindhyan sandstone and alluvium also have a higher capacity for stream development. The joints on Vindhyan Sandstone develop short fingertip streams which give a comparatively higher drainage density. The alluvium consisting mostly of clays and loams also develop short streams in the form of gullies which are usually denoted by the term 'badland topography'. Apart from the significant role played by underlying lithology semi-arid nature of climate also plays

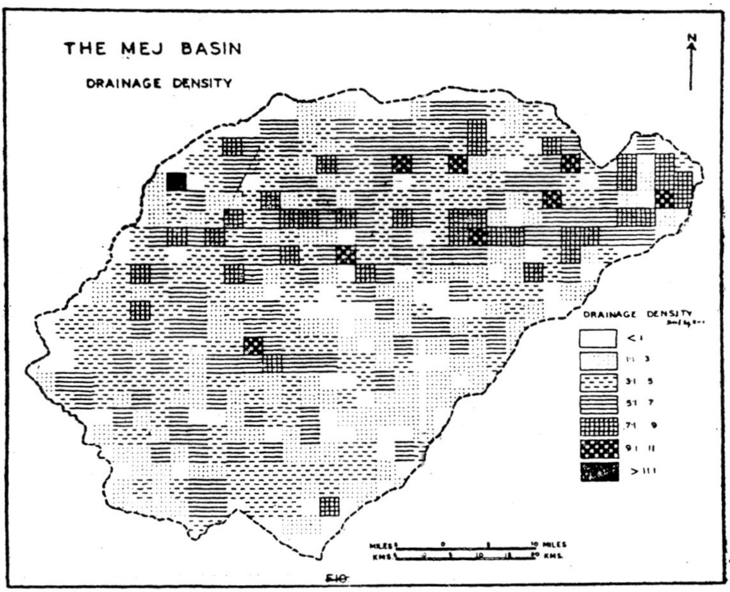

FIG. 7.14 : *The Mej Basin—drainage density*

its own part in increasing the intensity of rains during monsoons in this region. Further, the sandstone because of its resistant character has low infiltration capacity and, therefore, the runoff is more in the form of streams. Drainage density is directly related to the amount and intensity of precipitation and inversely to the amount of vegetation cover (Chorley, 1957). The less vegetation on hill slopes keeps to maintain the runoff rate.

TABLE 7.5 : **Drainage density in different lithological groups**

Group 1	Group 2	Group 3	Group 4	Group 5
.69	.37	.84	.51	.39
.42	.53	.87	—	.708
.69	.52	1.13	—	.508
.86	—	.59	—	.49
.507	—	.71	—	.34
.97	—	—	—	—

The Table 7.5 also shows higher values of drainage densities for more or less all the basins in group one and three and for basin 17 in group five.

Stream frequency

Stream frequency is the number of streams per square unit (Fig. 7.15) which along with drainage density gives the character of underlying lithology in a particular area. Stream frequency has the highest value in basins 4 and 12 depicting a case, similar to that of drainage density. These two basins fall again in Vindhyan sandstone and Gwaliors groups. The values for these two basins being 2.63 and 4.37 respectively and the minimum value of .32 falls again in basin 20 which has no doubt the least number of streams. Rest of the basins have a value around 1.5. Average values are also of the same pattern, with group three having a value of 2.34 and the group first has a value of 1.91. The Table 7.6 gives the values for all the basins in different types of lithologies.

TABLE 7.6 : **Stream frequencies in various lithological groups**

Group 1	Group 2	Group 3	Group 4	Group 5
2.39	.46	1.85	.64	.56
.75	.86	2.27	—	1.51
1.98	.41	4.37	—	.902
2.63	—	2.18	—	.43
1.01	—	1.04	—	.32
2.70	—	—	—	—

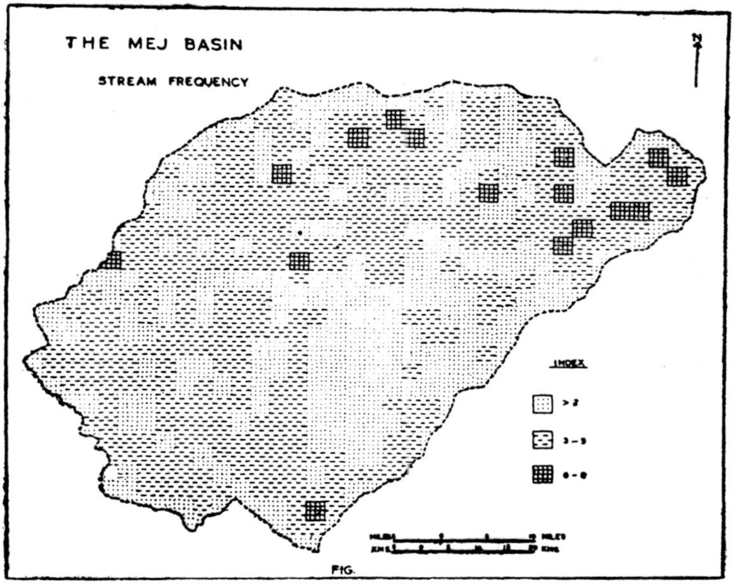

FIG. 7.15 : *The Mej Basin—Stream frequency*

The Table 7.6 also shows similar results to that of drainage density. Groups one and three have the highest stream frequencies along with basin 17 in group five, the causes of which are similar to that of drainage density.

Texture ratio

Texture ratio is number of contour crenulation by the length of perimeter of the basin. The values of texture ratio could be obtained only for 10 basins, as the contour values of these basins were only available from 1 : 50,000 scale toposheets. These ten basins belong to groups one, three and four. The values of texture ratios for these ten basins are tabulated below :

TABLE 7.7 : **Values of texture ratios in different lithologies**

Group 1	Group 2	Group 3	Group 4	Group 5
3.5	—	1.6	1.3	—
.41	—	.77	—	—
2.66	—	1.37	—	—
1.6	—	2.25	—	—
.73	—	—	—	—

The Table 7.7 though handicapped by the absence of other values still gives a generalised picture, resembling that of drainage density and stream frequency. The values are once again higher in groups one and three which is highly significant. The highest value of 3.5 falls in basin 1 which can be wholly categorised under the Gwaliors. In this particular basin, the number of crenulations being more the length of the perimeter is less, thereby gives a higher value. Strahler's (1952) concept of more number of streams with more number of crenulations fail here. Though the basin has more number of crenulations but the streams per unit are comparatively lesser.

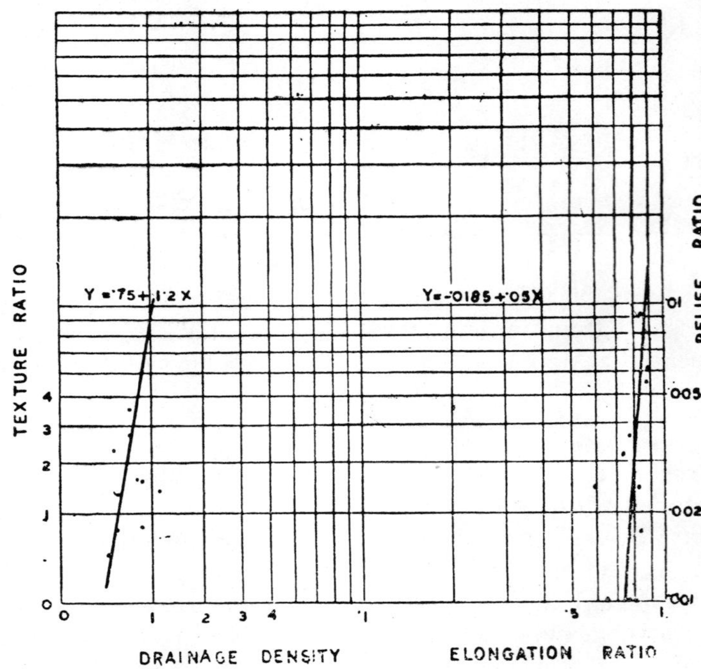

FIG. 7.16 : *Drainage density v/s Texture ratio*

FIG. 7.17 : *Elongation ratio v/s Relief ratio*

On a log^2 paper texture ratio is plotted against drainage density (Fig. 7.16) showing a positive exponential relation between the two parameters. The equation of the fitted line is :

$$Y = .75 + 1.2X$$

As Strahler (1954) has stated, drainage density which is a function of surface resistivity, runoff intensity and slope is the most sensitive indicator of the factors controlling the erosional development and has a close relation with the texture ratio. Both these factors are directly related with each other and speak of the erosional development of fingertip streams in the Gwaliors and the Vindhyan sandstone and alluvium.

Elongation ratio

According to Schumm (1956) the shape of any drainage basin is expressed by elongation ratio, the ratio between the diameter of a circle with the same area as the basin and the maximum length of the basin. It is clearly evidenced from Table IX that in all the basins the value of elongation ratio approaches 1.

TABLE 7.8 : **Values of elongation ratio in different basins**

Group 1	Group 2	Group 3	Group 4	Group 5
.88	.53	.87	.82	.62
.84	.72	.72	—	.79
.76	.53	.76	—	.85
.82	—	.58	—	.65
.64	—	.99	—	.83
.78	—	—	—	—

The basins falling in groups one and three have the highest elongation ratios. According to the different values of elongation ratio following shapes of basins have been obtained as has been suggested by geomorphologists.

TABLE 7.9 **Basin shape**

Basin shape	Circular	Elongated	Oval
The distinguishing value of elongation ratio.	.66	Around .57	.6 – .66
Number of basins	1, 2, 3, 4, 6, 8, 10, 11, 12, 14, 15, 17, 18, 20	9, 7, 13	5, 16

It can be noted from the Table 7.9 that as the basin shape increases, it becomes elongated in shape.

Circularity ratio

Similar to elongation ratio Miller (1953) used a similar measure, which is the ratio of circumference of a circle with same area, as the basin to basin perimeter. These values too range around 1. The highest value in 1.35 falls in basin 6 and secondly has a value of 1.15 in basin 14 as the areas of these two basins are the largest compared to other basins.

TABLE 7.10 : **A general picture of the values of circularity ratio**

Group 1	Group 2	Group 3	Group 4	Group 5
.65	.95	.77	.71	.87
.54	.47	.60	—	.56
.58	.73	.65	—	.99
.48	—	1.04	—	.46
.79	—	1.15	—	.79
1.35	—	—	—	—

Group three leads the rest of the groups in having higher values of circularity ratio. The areas of the basins falling in this group are larger as they contain larger number of streams.

Relief ratio

Schumm (1956) has defined relief ratio as the ratio between the total relief of a basin and the longest dimension of the basin parallel to principle drainage line. This is a dimensionless height length ratio. Relief ratio has the highest value in basin 10 of group third, the value of which is .054. Once again the values of ten basins could only be calculated. The Table 7.11 depicts some of the relief ratio values.

TABLE 7.11 : **Some of the relief ratio values**

Group 1	Group 2	Group 3	Group 4	Group 5
.006	—	.054	.024	—
.009	—	.031	—	—
.001	—	.036	—	—
.017	—	.024	—	—
.001	—	—	—	—

The values are naturally higher in the basins of group third because of the fact that the basins fall in a zone of contrasting relief features. Along with the relative relief values, the basins also have comparatively higher lengths. But basin 13 is exceptionally elongated. This is because of the fact that this basin has developed along the alternate beds of weak and hard strata, thereby the basin has a more chance of lengthening than widening. As has been referred earlier in the hypsometric analysis, the same basins falls under equilibrium stage in terms of stage of development where maximum relief predominates.

The values of relief ratio were plotted against elongation ratio on a logarithmic paper and the fitted line shows a relation of positive exponential function, where both the parameters increase in direct proportion (Fig. 7.17).

The equation for the fitted curve is as follows :

$$Y = -.0185 + .05X$$

Schumm (1956) has stated that as the relief ratio increases, the drainage basin becomes more elongated, which can be noticed from the above function of the fitted line.

Constant of channel maintenance

Constant of channel maintenance is a measure of texture similar to drainage density. This is inverse of drainage density and is always expressed in square kilometres. These values, in contrast to above results, have comparatively higher values in group five which falls in alluvium group. The highest value of 2.94 belongs to basin 20, where the drainage density is the lowest. The least value of .84 falls in group three which has the highest density. This analysis clearly shows that the intricate channels that have developed in basin 12 need less space to flow over than the channels in the alluvium.

TABLE 7.12 : **The values of different basins shows the generally prevailing higher values in group five.**

Group 1	Group 2	Group 3	Group 4	Group 5
1.45	2.70	1.19	1.96	2.59
2.38	1.89	1.15	—	1.41
1.45	1.92	.84	—	1.97
1.16	—	1.7	—	2.04
1.97	—	1.408	—	2.94
1.03	—	—	—	—

Average length of overland flow

Along with constant of channel maintenance length of overland flow is also an important factor in measuring the texture. Horton (1945) calculated the length of overland flow as the reciprocal of twice the drainage density. This has also an effective relation with drainage density and constant of channel maintenance. The results of this parameter also lie close to that of channel maintenance. These values are just half the values of constant of channel maintenance, which can be noted from Table 7.13.

TABLE 7.13 : **Constant of channel maintenance**

Group 1	Group 2	Group 3	Group 4	Group 5
.72	1.35	.595	.980	1.295
1.19	.945	.575	—	.705
.72	.96	.420	—	.985
.58	—	.850	—	1.02
.985	—	.704	—	1.47
.515	—	—	—	—

The averages too are of same behaviour, the highest value of 1.95 falls under the group of alluvium and a value of 1.8 falls under the category of Vindhyan sandstone, which is of similar nature as the previous result. The least value of .629 once again falls under Vindhyan sandstone and alluvium group. These results clearly show that the streams developing on Vindhyan lithology are short, joint controlled, whereas those developing on alluvium are longer in their courses because of the case of headward erosion in this type of soils. The same result also leads to the fact that the spacing of streams is less on the Vindhyans, whereas wider in alluvium. The streams on alluvium flow longer lengths before draining into the main channel.

Basin length

Basin length is the longest dimension of the basin, parallel to principle drainage line. These values indicate the shape and size of the basin along with basin shape.

TABLE 7.14 : **The values of basin lengths in different groups**

Length in kilometres

Group 1	Group 2	Group 3	Group 4	Group 5
6.1	15.56	5.5	5.45	10.84
7.1	8.38	5.5	—	5.10
4.85	12.29	6.3	—	8.51
3.85	—	18.40	—	17.45
9.15	—	7.94	—	7.94
14.05	—	—	—	—

The above values indicate lower values for group one and in contrast, extraordinarily high values for group five. The basins falling under the alluvium group have the largest lengths mainly because of the fact that the basins are larger in area, as to support longer streams that have developed due to headward erosion of the streams. Basin 13 of group three also has the highest value of 18.40 due to elongated nature of the basin, that has developed on alternate bands of soft and hard rocks.

SUGGESTED READINGS

Chorley, R. J., 1958 : Climate and morphometry, *Journal of Geology*, Vol. 65, pp. 628-638.

Chorley, R. J., 1966 : The application of statistical methods to geomorphology, *Essays in Geomorphology*, American Elsevier Publishing Company, New York, pp. 275-387.

Clarke, J. I., 1966 : Morphometry from maps, *Essays in Geomorphology*, American Elsevier Publishing Company, New York, pp. 235-274.

Ghose, B.; Pandey; Singh and Gheesa Lal, 1967 : Quantitative geomorphology of the drainage basins in the central Luni Basin in Western Rajasthan, *Z. Geomorph. N.F.*, Vol. 11, pp. 146-160.

Horton, R. E., 1945 : Erosional development of streams and their drainage basins, hydrophysical approach to quantitative morphology, *Bulletin of the Geological Society of America*, Vol. 56, pp. 275-370.

Kesseli, J. F., 1946 : A neglected field of geomorphology, *Annals of Association of American Geographers*, Vol. 86, p. 93.

Leopold, Wolman, Miller, 1964 : *Fluvial Processes Geomorphology Eurasis*, Publishing House Private Limited, New Delhi.

Melton, M. A., 1958 : Correlation structure of morphometric properties of drainage systems and their controlling agents, *Journal of Geology*, Vol. 66, pp. 442-460.

Miller, V. C., 1953 : Relation of quantitative geomorphic study of drainage basin characteristics in the Clinch Mountain Area, *Virginia and*

Tennesse, Technical Report 3, Office of the Naval Research, Department of Geology, Columbia University, New York.

Morisawa, M. E., 1962 : Relation of quantitative geomorphology of streams flow in representative watersheds of Appalachian plateau province, *Bulletin of Geological Society of America*, Vol. 73, pp. 1025-1046.

Pal, S. K., 1973 : Quantitative geomorphology of drainage basins of Himalayas, *Geographical review of India*, Vol. 35, pp. 81-101.

Robinson, G,, 1963 : A consideration of the relation of geomorphology and geography, *Professional Geographer*, Vol. 12, No. 2, pp. 5-15.

Russell, R. J., 1949 : Geographical geomorphology, *Annals of Association of American Geographers*, Vol. 39, p. 10.

Sharma, V., 1968 : Quantitative terrain types in Chaur-Rajgarh tract of the lesser Himalayas, *Geographical Review of India*, Vol. 30, No. 1. pp. 31-42.

Schumm, S. A., 1956 : Evolution of drainage systems and slopes in badlands at Perth Amboy, New Jersey, *Bulletin of Geological Society of America*, Vol. 67, pp. 597-646.

Smith, K. G., 1958 : Erosional processes and landforms in badlands, National Monument, S. Dacota, *Bulletin of Geological Society of America*, Vol. 69, pp. 975-1008.

Strahler, A. N., 1950 : Equilibrium theory of erosional slopes approached by frequency distribution analysis, *American Journal of Science*, Vol. 248, pp. 673-696.

Strahler, A. N., 1952 : Hypsometric (Area Altitude) analysis of erosional topography, *Bulletin of Geological Society of America*, Vol. 63, pp. 1117-42.

Strahler, A. N., 1952 : Quantitative geomorphology of erosional landscapes, 19th International Geological Congress, *Algiers, Section 3*, Vol. 3, pp. 341-354.

Strahler, A. N., 1953 : Revisions of Horton's quantitative factors in erosional terrain, Paper read before Hydrology Section of American Geographical Union, Washington, D. C.

Strahler, A. N., 1954 : Statistical analysis in geomorphic research, *Journal of Geology*, Vol. 62, pp. 1-25.

Strahler, A. N., 1958 : Dimensional analysis applied to fluvially eroded landforms, *Bulletin of Geological Society of America*, Vol. 64, pp. 279-300.

Strahler, A. N., 1971 : Quantitative analysis of erosional landforms, *Physical Geography*, Wiley Eastern Private Limited in New Delhi, pp. 481-500.

Tandon, S. K., 1974 : Litho-Control of some geomorphic properties : An illustration from Kumaun Himalayas, India, *Z. Geomorph. N. F.*, Vol. 18, pp. 460-71.

Verma, V. K. and Tandon, S. K., 1971 : Stream tortuosity, a new method of visual estimation, *Z. Geomph. N. E.*, Vol. 12, pp. 165-71.

APPENDIX I

QUANTITATIVE GEOMORPHIC CHARACTERISTICS OF THE STREAMS OF THE MEJ BASIN

Basin Number	Basin Order	Number of streams	B.R.	Weighted mean B.R.S.	T.S.L. of order	Cumulative (T.S.L.) (Kms.)	C.R. (T.S.L.) (Kms.)	Weighted mean C.R. (T.S.L.) (Kms.)
1	2	3	4	5	6	7	8	9
1	1	30	4.29		6.750	6.750	.31	
	2	7	3.50	4.004	2.125	8.875	.65	.52
	3	2	2.00		1.375	10.250	1.00	
	4	1			1.375	11.625		
2	1	15	3.00		5.5	5.5	.77	
	2	5	5.00	3.46	4.25	9.75	.50	.66
	3	1			2.125	11.875		
3	1	17	4.25		4.375	4.375	.26	
	2	4	4.00	4.20	1.125	5.500	2.00	.92
	3	1			2.250	7.750		
4	1	15	3.00		3.75	3.75	.33	
	2	5	5.00	3.46	1.25	5.00	1.50	.78
	3	1			1.875	6.875		

(contd.)

Appendix I—(contd.)

Basin Number	Basin Order	M.S.L. (Kms)	Cumulative (M.S.L.) (Kms.)	C.R. (M.S.L.) (Kms.)	Weighted C.R. (M.S.L.)	Area (T.S.L.) of orders (Sq. Kms.)	A_u	Cumulative basin area	Lb (Kms.)
1	2	10	11	12	13	14	15	16	17
1	1	.225	.225	1.4		9.7	.31	9.7	
	2	.304	.529	2.3	1.9	3.05	.62	12.75	6.1
	3	.688	1.217	2.00		1.9	1.0	14.65	
	4	1.375	2.592			1.9		16.55	
2	1	.370	.370	2.30		12.9	.77	12.80	
	2	.850	1.220	2.50	3.16	9.9	.89	22.70	7.1
	3	2.125	3.345			4.9		27.60	
3	1	.257	.257	1.09		6.25	.26	6.25	
	2	.280	.532	8.03	6.8	1.60	1.00	7.85	4.85
	3	32.250	2 787			3.03		10.88	
4	1	.250	.250	1.00		4.30	.32	4.30	
	2	2.250	.50	7.5	6.2	1.40	1.4	5.70	3.85
	3	1.875	2.375			2.02		7.72	

Basin Number	Basin Order	Drainage density	Stream frequency	Elongation ratio	Circulatory ratio	Basin shape	Average length of overflows (Kms.)	Texture ratio	Relief ratio
1	2	18	19	20	21	22	23	24	25
1	1								
	2	.69	2.39	.88	.65	.51	.72	3.5	.006
	3								
	4								
2	1								
	2	.42	.75	.84	.54	.55	1.19	.41	.009
	3								
3	1								
	2	.69	1.98	.76	.58	.47	.72	2.66	.001
	3								
4	1								
	2	.86	2.63	.82	.48	.53	.58	1.6	.017
	3								

(*contd.*)

Appendix I—(contd.)

Basin Number	Basin Order	Channel maintenance	Basin area (M.S.L.)	A_u	Cumulative basin area	T.S.L. (Kms.)	M.S.L. (Kms.)	Total basin area (Sq. Kms.)	Total Number of Streams
1	2	26	27	28	29	30	31	32	33
1	1		1.45	1.3	1.45				
	2	1.45	1.90	2.3	3.35	11.625	.291	16.69	40
	3		4.40	2.0	7.75				
	4		8.80		16.55				
2	1		3.07	2.3	3.07				
	2	2.38	7.07	2.5	10.14	11.875	.565	27.82	21
	3		17.69		27.83				
3	1		1.02	1.09	1.02				
	2	1.45	1.11	8.01	2.13	7.750	.352	11.08	22
	3		8.90		11.03				
4	1		.83	1.00	.83				
	2	1.16	.83	7.4	1.66	6.875	.327	7.96	21
	3		6.2		7.86				

1	2	3	4	5	6	7	8	9
5	1	23	5.75		6.50	6.50	.46	
	2	4	4.00	3.54	3.00	9.50	1.50	.91
	3	1			4.50	14.00		
6	1	247	4.20		17.82	17.82	.22	
	2	59	3.30	3.96	15.75	87.57	1.20	.49
	3	18	3.60		18.90	106.47	.26	
	4	5	2.50		5.04	111.51	1.00	
	5	2	2.00		5.04	116.55	.50	
	6	1			2.52	119.07		
7	1	26	5.20		15.120	15.120	.43	
	2	5	5.00	5.17	6.615	21.735	.66	.50
	3	1			4.410	26.145		
8	1	22	3.14		10.710	10.710	.44	
	2	7	3.50	3.14	4.725	15.435	.73	.47
	3	2	2.00		3.465	18.900	.09	
	4	1			.315	19.215		

(contd.)

Appendix I—(contd.)

1	2	10	11	12	13	14	15	16	17
5	1	.280	.280	2.67		12.81	.49	12.81	
	2	.750	.93	6.00	5.4	5.91	1.9	18.71	9.15
	3	4.50	5.43			8.87		27.58	
6	1	.29	.29	.91		73.52	.13	73.52	
	2	.265	.555	3.90	1.69	10.01	1.9	83.53	
	3	1.05	1.605	.96		19.34	.26	102.87	14.05
	4	1.01	1.615	.49		5.16	1.0	108.63	
	5	2.52	5.135	1.00		5.16	.50	113.19	
	6	2.52	7.655			2.58		115.77	
7	1	.580	.580	2.28		42.7	.40	42.7	
	2	1•325	1.905	3.32	3.06	17.43	.66	60.13	15.56
	3	4.410	6.315			11.62		71.75	
8	1	.485	.485	1.39		20.10	.44	20.10	
	2	.675	1.160	2.56	1.50	8.89	.73	28.99	8.38
	3	1.730	2.890	.18		6.52	.09	35.51	
	4	.315	3.205			.59		36.10	

1	2	18	19	20	21	22	23	24	25
5	1								
	2	.507	1.01	.64	.79	.32	.985	.73	.001
	3								
6	1								
	2	.97	2.70	.78	1.35	.62	.515		
	3								
	4								
	5								
	6								
7	1								
	2	.37	.46	.53	.95	.28	1.35		
	3								
8	1								
	2	.53	.86	.72	.47	.51	.945		
	3								
	4								

(contd.)

Appendix I—(contd.)

1	2	26	27	28	29	30	31	32	33
5	1		1.4	2.7	1.4				
	2	1.97	3.8	6.0	5.2	14.000	.500	27.60	28
	3		22.8		28.0				
6	1		4.6	.91	4.6				
	2	1.03	4.2	4.0	8.8	119.070	.358	121.90	332
	3		16.7	.95	25.5				
	4		16.5	2.40	42.0				
	5		40.13	1.00	82.13				
	6		40.13		122.26				
7	1		6.3	2.3	6.3				
	2	2.70	14.4	3.3	20.7	26.145	.817	68.92	32
	3		48.1		68.8				
8	1		5.6	1.4	5.6				
	2	1.89	7.8	2.4	13.4	19.215	.600	36.16	32
	3		19.1	.17	32.5				
	4		3.3		35.8				

1	2	3	4	5	6	7	8	9
9	1	13	2.60		13.860	13.860	.068	
	2	5	5.00	3.20	.945	14.805	9.60	3.7
	3	1			9.135	23.940		
10	1	25	4.17		9.750	9.750	307	
	2	6	3.00	3.79	3.000	12.750	54	.41
	3	2	2.00		1.625	14.375	.68	
	4	1			1.125	15.500		
11	1	23	5.75		5.10	5.10	.74	
	2	4	4.00	5.48	3.75	8.85	.50	.64
	3	1			1.875	10.725		
12	1	63	4.50		12.75	12.75	.29	
	2	14	4.67	4.47	3.75	16.50	.73	.45
	3	3	3.00		2.75	19.25	.64	
	4	1			1.75	21.00		

(*contd.*)

Appendix I–(contd.)

1	2	10	11	12	13	14	15	16	17
9	1	1.065	1.065	.18		26.48	.07	26.48	
	2	.190	1.255	48.08	42.4	1.90	9.2	28.38	12.29
	3	9.135	10.390			17.48		45 86	
10	1	.395	.395	1.26		11.53	.31	11.53	
	2	.500	.895	1.62	1.4	3.55	.55	15.08	5.5
	3	.812	1.608	1.38		1.91	.69	16.99	
	4	1.125	2.733			1.33		18.32	
11	1	.230	.230	4.08		5.85	.73	5.85	
	2	.937	1.167	2.00	2.6	4.30	.50	10.15	5.5
	3	1.875	3.043			2.15		12.30	
12	1	.20	.20	1.1		11.25	.29	11.25	
	2	.22	.42	4.09	2.4	3.31	.73	14.56	6.3
	3	.90	1.32	1.96		2.42	.63	16.98	
	4	1.75	3.07			1.54		17.52	

1	2	18	19	20	21	22	23	24	25
9	1	.52	.41	.53	.73	.30	.96		
	2								
	3								
10	1	.84	1.85	.87	.77	.60	.595	1.6	.054
	2								
	3								
	4								
11	1	.87	2.27	.72	.60	.40	.575	.77	.031
	2								
	3								
12	1	1.13	4.37	.76	.65	.46	.420	1.37	.036
	2								
	3								
	4								

(*contd.*)

Appendix I—(contd.)

1	2	26	27	28	29	30	31	32	33
9	1		4.4	.20	4.4				
	2	1.92	.9	44.4	5.3	23.940	1.260	45.84	19
	3		40.0		45.3				
10	1		2.7	1.2	2.7				
	2	1.19	3.35	1.6	6.05	15.500	.456	18.34	34
	3		5.36	1.4	11.41				
	4		7.30		18.71				
11	1		.8	4.5	.8				
	2	1.15	3.6	2.1	4.4	10.725	.383	12.31	28
	3		7.6		12.0				
12	1		1.20	1.1	1.20				
	2	.84	1.32	4.1	2.52	21.000	.260	18.53	81
	3		5.43	1.8	7.95				
	4		9.80		17.75				

(contd)

1	2	3	4	5	6	7	8	9
13	1	137	2.68	3.97	24.75	24.75	.65	.58
	2	51	8.50		16.25	41.00	.307	
	3	6	3.00		5.00	46.00	.80	
	4	2	2.00		4.00	50.00	.75	
	5	1			3.00	53.00		
14	1	51	4.25	4.47	28.35	28.35	.30	.41
	2	12	6.00		8.505	36.855	.81	
	3	2	2.00		6.930	43.785	.18	
	4	1			1.260	45.045		
15	1	7	3.50	3.125	4.000	4.000	.906	.63
	2	2	2.00		3.625	7.625	.41	
	3	1			.500	8.125		
16	1	18	3.60	3.22	8.820	8.820	.36	.71
	2	5	2.50		3.150	11.970	1.60	
	3	2	2.00		5.04	17.010	.21	
	4	1			1.07	18.080		

(contd.)

Appendix I—(contd.)

1	2	10	11	12	13	14	15	16	17
13	1	.18	.18	1.77		42.02	.65	42.02	
	2	.32	.50	2.60	1.1	27.59	.30	69.61	18.40
	3	.835	1.335	2.39		8.49	.79	78.10	
	4	2.000	3.335	1.50		6.79	.75	84.89	
	5	3.000	6.335			5.09		89.98	
14	1	.555	.555	1.27		39.66	.30	39.66	
	2	.705	1.260	4.90	2.3	11.90	.81	51.56	7.94
	3	3.465	4.725	.32		9.69	.18	61.25	
	4	1.260	5.985			1.79		63.01	
15	1	.570	.570	3.28		7.81	.89	7.81	
	2	1.815	2.385	.27	1.79	7.00	.14	14.81	5.45
	3	.500	2.885			.97		15.78	
16	1	.49	.49	1.28		23.35	.35	22.35	
	2	.63	1,12	4.00	1.9	7.9	1.5	30.25	10.84
	3	2.52	3.64	.42		12.00	.22	42.47	
	4	1.07	4.71			2.70		45.17	

1	2	18	19	20	21	22	23	24	25
13	1								
	2	.59	2.18	.58	1.04	.27	.850	2.25	.024
	3								
	4								
	5								
14	1								
	2	.71	1.04	.99	1.15	1.01	.704		
	3								
	4								
15	1								
	2	.51	.64	.82	.71	.54	.980	1.3	.024
	3								
16	1								
	2	.39	.56	.62	.87	.39	1.295		
	3								
	4								

(contd.)

Appendix I—(contd.)

1	2	26	27	28	29	30	31	32	33
13	1		2.80	1.5	2.8				
	2	1.7	4.30	2.6	7.1	53.00	.270	90.01	197
	3		11.30	2.4	18.04				
	4		28.40	1.5	46.8				
	5		42.60		89.4				
14	1		6.30	1.15	6.30				
	2	1.408	7.30	5.04	13.6	45.045	.682	63.03	66
	3		36.80	.28	50.4				
	4		10.50		60.9				
15	1		3.20	3.06	3.2				
	2	1.96	9.80	.27	13.0	8.125	.8125	15.69	10
	3		2.70		15.7				
16	1		4.80	1.2	4.8				
	2	2.59	5.80	4.2	10.6	18.080	.695	45.84	26
	3		24.30	.39	34.9				
	4		9.70		44.6				

1	2	3	4	5	6	7	8	9
17	1	20	6.67		8.820	8.820	.32	
	2	3	3.00	6.01	2.835	11.655	.22	.29
	3	1			.630	12.285		
18	1	37	5.30		15.75	15.75	.52	
	2	7	3.50	4 83	8.19	23.94	.23	.42
	3	2	2.00		1.89	25.83	.33	
	4	1			.63	26.46		
19	1	33	4.1		28.98	28.98	.69	
	2	8	4.0	3.96	20.16	49.14	.46	.57
	3	2	2.0		9.45	58.59	.40	
	4	1			3.78	62.37		
20	1	7	2.33		7 560	7.560	.29	
	2	3	3.00	2 52	2.205	9.765	.86	.45
	3	1			1.890	11.655		

(contd.)

Appendix I—(contd.)

1	2	10	11	12	13	14	15	16	17
17	1	.440	.440	2.15		11.36	.32	11.36	
	2	.945	1.385	.66	1.40	3.65	.22	15.01	5.10
	3	.630	2.015			.81		15.82	
18	1	.425	.425	2.75		31.00	.52	31.00	
	2	1.170	1.595	.807	1.3	16.12	.23	47.12	8.51
	3	.945	2.540	.66		3.67	.33	50.79	
	4	.630	3.170			1.24		52.03	
19	1	.88	.88	2.86		59.2	.35	59.2	
	2	2.520	3.400	1.79	1.4	20.77	.93	79.97	17.45
	3	4.725	8.125	.50		19.32	.40	99.29	
	4	3.780	11.905			7.72		107.21	
20	1	1.080	1.080	.68		22.50	.28	22.50	
	2	.735	1.815	2.57	1.8	6.35	.71	28.85	7.94
	3	1.890	3.705			5.45		34.30	

1	2	18	19	20	21	22	23	24	25
17	1								
	2	.708	1.51	.79	.56	.61	.705		
	3								
18	1								
	2	.508	.902	.85	.99	.72	.985		
	3								
	5								
19	1								
	2	.49	.43	.65	.46	.41	1.02		
	3								
	4								
20	1								
	2	.34	.32	.83	.79	.54	1.47		
	3								

(contd.)

Appendix I—(contd).

1	2	26	27	28	29	30	31	32	33
17	1		3.14	.23	3.14				
	2	11.41	7.4	.66	10.54	12.285	.512	15.85	24
	3		4.9		15.44				
18	1		6.5	3.03	6.5				
	2	1.97	19.7	.78	26.2	26.460	.563	52.09	47
	3		15.5	.63	41.7				
	4		9.8		51.5				
19	1		9.6	2.8	9.6				
	2	2.04	26.7	1.9	36.3	62.370	1.417	127.51	44
	3		50.3	.80	86.6				
	4		40.7		127.3				
20	1		9.1	.72	9.1				
	2	2.94	6.6	2.6	15.7	11.755	1.068	33.89	11
	3		17.4		33.1				

Abbreviations : B.R. = Bifurcation ratio
T.S.L. = Total stream length
C.R. = Channel ratio
M.S.L. = Mean stream length
Lb = Basin length
A_u = Area ratio

APPENDIX II

AVERAGE QUANTITATIVE GEOMORPHIC CHARACTERISTICS OF THE MEJ BASIN IN DIFFERENT LITHOLOGICAL GROUPS

Parameter	Gwaliors 1,2,3,4,5,6	Upper Bhander Vindhyan Sandstone 7,8,9	Vindhyan Sandstone Limestone Alluvium 10,11,12,13,14	Gwaliors Kaimur Sandstone Alluvium 15	Alluvium 16,17,18,19,20
1	2	3	4	5	6
1. Total number of streams	21.09	8.3	20.3	3.3	8.4
2. Total stream length	7.78	6.93	7.26	2.70	7.72
3. Mean stream length	.108	.267	.102	.270	.236
4. Bifurcation ratio	3.4	3.77	3.96	2.75	3.38
5. Weighted mean bifurcation ratio	3.77	3.83	4.43	3.125	4.11
6. Drainage density	.69	.47	.83	.51	.48
7. Stream frequency	1.91	.58	2.34	.64	.74
8. Elongation ratio	.78	.59	.78	.82	.75
9. Circularity ratio	.73	.71	.84	.71	.73
10. Basin area from TSL	9.37	15.31	10.10	5.26	14.13
11. Area ratio (A_u) from TSL	.85	1.65	.54	.51	.49
12. Area ratio from MSL	2.86	2.02	2.15	1.66	1.53

(contd.)

Appendix II—(contd.)

1	2	3	4	5	6
13. Channel ratio from total stream length	.78	1.71	.615	.52	.49
14. Channel ratio from mean stream length	2.89	8.28	2.15	1.77	1.625
15. Weighted total channel ratio (TSL)	.71	1.56	.498	.63	.488
16. Weighted mean channel ratio (MSL)	4.19	15.65	1.96	1.79	1.56
17. Basin shape	.50	.36	.55	.54	.53
18. Channel maintenance	1.57	2.17	1.257	1.960	2.19
19. Average length of overland flow	.962	1.8	.629	.980	1.95
20. Basin area from MSL	9.7	14.99	9.94	5.23	15.10
21. Cumulative total stream length	34.58	17.98	25.76	6.58	22.54
22. Cumulative mean stream length	1.85	2.92	1.97	1.93	2.91
23. Cumulative area from total stream length	37.97	39.60	36.53	12.80	44.14
24. Cumulative area from mean stream length	18.84	23.81	18.56	10.30	31.53
25. Length of Basin (Lb)	7.52	12.08	8.73	15.69	55.03
26. Basin area (whole)	35.5	50.3	40.4	15.69	55.03

Abbreviations : TSL=Total stream length
MSL=Mean stream length

EIGHT

Savindra Singh and *D. P. Upadhyaya*

TOPOLOGICAL AND GEOMETRIC STUDY OF DRAINAGE NETWORK, S.E. CHOTA NAGPUR REGION (INDIA)

PROLOGUE

THE drainage basin accepted as an ideal areal unit for the interpretation and analysis of fluvially originated landforms exhibits an example of open system of system operation wherein its various components viz. stream segments, basin length, basin perimeter and basin area experience orderly growth in terms of law of allometric growth indicating dynamic nature of basin development, but tectonic, biotic and climatic factors introduce anomalies in its natural and systematic growth. Keeping these facts in view, 30 basins of the 6th and 5th order of different physiographic regions of the S.E. Chota Nagpur region having their own distinctive structural, tectonic, biotic and climatic characteristics have been selected to study the nature of development of open and closed links of drainage network.

The region

The S.E. Chota Nagpur region (21° 58′ N-23° 11′ N, 85°-86° 54′ E), spreading over entire Singhbhum district of Bihar and covering an area of 13,447 sq. kms., represents the south-eastern segment of Chota Nagpur Highland region. The region is characterized by different geological formations viz. Dharwarian and Cuddapah formations in the western highlands, granitic-gneissic structure over the central plateau, conglomer-ate, phyllites and lavas of Dalma and Dhanjori highlands and formations of the Iron-Ore-Series overlain by Tertiary grits, gravels and recent alluvium of the Subarnarekha plain. The region has experienced several phases of folding (Archaean folding of Dharwarian sedimentaries), faulting, vulcanicity, tilting, upliftment (three phases of Tertiary Orogeny), erosion and deposition resulting into the formation of 'palimpsests'. The annual rainfall varies from 1200 mm to over 1500 mm and the vegetal cover includes dry deciduous, moist deciduous and shrubs and grasses with varying density over different physiographic regions. These two elements have played major roles in the process of pedogenesis so as to give birth to laterite soil, eroded escarp soil, red soil, gravel soil and alluvial soils. The distinctive sets of climatic elements, plant cover and soils over the study region have produced definite micro-morphogenetic regions and not only have affected the morphogenetic processes but have also conditioned the network characteristics. The superimposition of different maps of average slopes, relative reliefs, dissection index, drainage density, stream frequency has resulted into the recognition of five physiographic regions viz. the western highland, the central plateau, the Dalma highland, the Dhanjori highland and Subarnarekha plain region.

Tentarigara, Koina, Kumirta, Sangjata, Sapu, Mahadeosalgara, Dadkagara, Kumbigara, Bodikuchagara, Roro, Ghagara Bamni and Banki basins drain over the western highlands; central plateau has been represented by Konkoa Kongera, Torlo, Karanjia, Gara III, Toyargara, Kuchai, Sona and Bhangbanga basins; Suru, Kantajuria, Kharsoti, Sapghara and Bagalduba basins are located over Dalma highlands but their lower reaches are spread over Subarnarekha plain and Sankh and Dudh basins represent Dhanjori highlands, Kuchai, Sasagara and Toyargara basins take their sources in the western highlands but their lower reaches are located over the central plateau region.

Segment hierarchy and stream number

The hierarchical orders of drainage network have been determined according to the ordinal scale of stream ordering presented by Strahler (1952, 1953, 1957 and in Chow's 1964) (Fig. 8.1). Table 8.1 denotes the fact that the stream number is independent of basin areas as is indicated by insignificant correlation

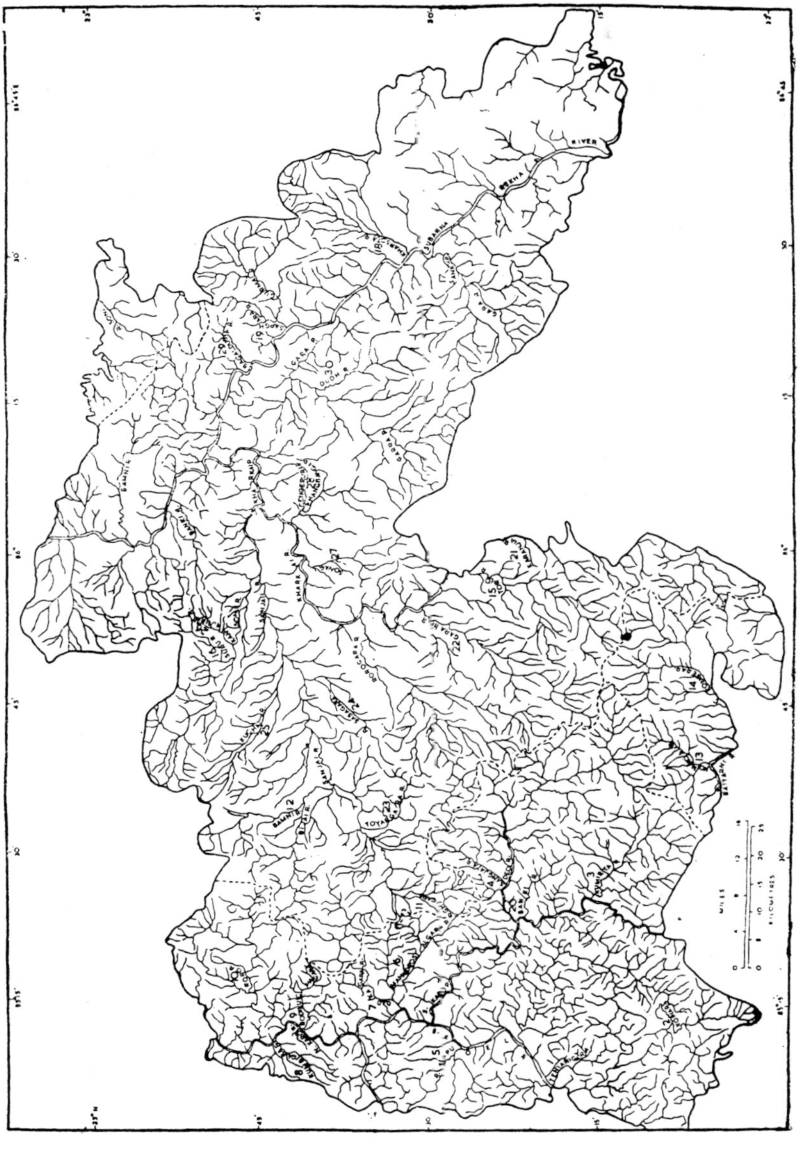

FIG. 8.1 ; *South-east Chota Nagpur region—Drainage and small drainage basins*

TABLE 8.1 : **Number of stream segments (N_u)**

Sl. No.	Basins	ORDER (U)						Total No. of stream segments
		N_{u1}	N_{u2}	N_{u3}	N_{u4}	N_{u5}	N_{u6}	
1	Tentarigara	579	112	26	7	2	1	727
2	Koina	3951	911	210	46	11	1	5130
3	Kumirta	1546	359	78	18	4	1	2006
4	Sangjata	742	184	34	6	2	1	969
5	Sapu	1240	288	59	14	3	1	1605
6	Mahadeosalgara	896	202	43	13	2	1	1157
7	Dadkagara	669	149	38	8	2	1	867
8	Kumhigara	594	127	29	7	2	1	760
9	Bodikuchagara	1109	238	44	11	3	1	1406
10	Roro	3694	817	167	40	9	1	4728
11	Ghagara	877	231	52	14	3	1	1178
12	Bamni	590	146	37	7	2	1	783
13	Konkoa	412	96	26	5	2	1	542
14	Kongera	788	170	42	8	2	1	1011
15	Torlo	750	175	43	10	3	1	982
16	Suru	631	125	35	7	2	1	801
17	Sankh	1516	287	59	12	3	1	1878
18	Kharsoti	534	115	28	6	2	1	686
19	Sapghara	945	218	51	13	4	1	1232
20	Banki	726	180	37	10	1	—	954
21	Karanjia	343	51	13	3	1	—	411
22	Gara III	527	130	24	6	1	—	688
23	Toyargara	424	111	18	5	1	—	559
24	Sasagara	159	36	9	2	1	—	207
25	Kuchai	405	91	26	5	1	—	528
26	Kantajuria	205	57	12	4	1	—	279
27	Sona	215	51	11	4	1	—	282
28	Bhangbanga	315	69	17	4	1	—	406
29	Bagalduba	140	28	8	3	1	—	180
30	Dudh	277	66	14	2	1	—	360

coefficient of +0.0006 between total number of stream segments and the area of 30 selected basins. It may, thus, be opined that lithological characteristics, pedological factors, slope, general relief, plant cover and degree of dissection have largely controlled the branching of the drainage basins. The weak Dharwarian formation of the western highlands together with high absolute reliefs, steep slopes, isolated hills and ranges and scarp faces have issued and maintained much feeders which are reflected in the larger number of stream segments of Koina, Kumirta, Roro, Sapu, Mahadeosalgara and Bodikuchagara basins. Higher degree of dissection of Dalma and Dhanjori highlands resulting into scarp faces of mesas and buttes and steep slopes therein has been responsible for the development of dense drainage network of Suru, Sankh, Kantajuria and Sapghara basins. The streams, draining over flattish and peneplained granitic-gneissic surface of the central plateau, have small number of stream segments with respect to drainage area because gentle slopes, comparatively hard rock formations, and low relative reliefs have restricted the branching of streams as is exemplified by Kongera, Konkoa, Torlo, Gara III, Sasagara and Toyaragara basins. The effect of structure and relief may be clearly observed in Kuchai and Bamni rivers which originate on the western highlands and after descending from the precipitous scarps they drain over the central plateau. In their upper reaches, the branching is well developed due to steep slopes, scarp faces, hills and tablelands but in their lower reaches, branching has been reduced because of level terrain of the central plateau. Similar is the case with Kharsoti and Sapghara basins which originate in the Dalma and phyllite ranges but after descending from these ranges they flow over the Subarnarekha plain. The mesas, dykes and sills of the Dalma highlands have presented ideal conditions for stream development in the upper reaches of the said rivers but the levelled surface of the Subarnarekha plain has restricted the development of stream segments.

Bifurcation ratio (R_b)

The average bifurcation ratios (R_b) of 27 drainage basins (Table II) range between 3.0 and 5.0 and, thus, the observation of Coates (1958) that 'minimum possible ratio of 2.00 is seldom approached in nature, lying between about 3.0 and 5.0 in basins without differential geological controls and only reaching higher values where geological controls favour the development of elongated narrow basins' is very much confirmed in the present study. Koina, Roro and Banki basins are exceptions of the above general rule having mean bifurcation ratios of 5.69, 5.41 and 5.64 respectively. This departure is the result of elongated shapes (circularity index being 0.37, 0.44 and 0.43 respectively) of these rivers because of the parallel ranges that guide the stream courses of their tributaries. Koina river lies between Budha Buru range and Sasangda and Chatkhuri ranges while Roro basin is located between the border ranges of the Ranchi plateau in the north and Lukud Buru and Kundrugutu ranges in the south. Similarly Banki basin is jagged between Matkam Buru range in the north and Latua range in the south.

TABLE 8.2 : **Bifurcation ratio (R_b)**

Sl. No.	Basins	N_1/N_2	N_2/N_3	N_3/N_4	N_4/N_5	N_5/N_6	R_b
1	Tentarigara	5.17	4.31	3.71	4.67	3.00	3.74
2	Koina	4.34	4.34	4.57	4.18	11.00	5.69
3	Kumirta	4.31	4.60	4.33	4.50	4.00	4.35
4	Sangjata	4.03	5.41	5.67	3.00	2.00	4.02
5	Sapu	4.31	4.88	4.21	4.67	3.00	4.21
6	Mahadeosalgara	4.44	4.70	3.31	6.50	2.00	3.74
7	Dadkagara	4.49	3.92	4.75	4.00	2.00	3.83
8	Kumbigara	4.68	4.38	4.14	3.50	2.00	4.19
9	Bodikuchagara	4.66	5.41	4.00	3.67	3.00	4.15
10	Roro	4.52	4.89	4.18	4.44	9.00	5.41
11	Ghagara	3.80	4.44	3.71	4.67	3.00	3.92
12	Bamni	4.04	3.95	5.29	3.50	2.00	3.76
13	Konkoa	4.29	3.69	5.20	2.50	2.00	3.54
14	Kongera	4.64	4.05	5.25	4.00	2.00	4.00
15	Torlo	4.28	4.07	4.30	3.33	3.00	3.80
16	Suru	5.05	3.57	5.00	3.50	2.00	3.82
17	Sankh	5.28	4.86	4.92	4.00	3.00	4.41
18	Kharsoti	4.64	4.11	4.67	3.00	2.00	3.68
19	Sapghara	4.33	4.27	3.92	3.25	4.00	3.93
20	Banki	4.03	4.86	3.70	10.00	—	5.64
21	Karanjia	4.76	3.92	4.33	3.00	—	4.00
22	Gara III	4.05	5.42	4.00	6.00	—	4.87
23	Toyargara	3.82	6.17	3.60	5.00	—	4.65
24	Sasagara	4.42	4.00	4.50	2.00	—	3.73
25	Kuchai	4.45	3.50	5.20	5.00	—	4.54
26	Kantajuria	3.59	4.75	3.00	4.00	—	3.84
27	Sona	4.22	4.64	2.75	4.00	—	3.90
28	Bhangbanga	4.57	4.06	4.25	4.00	—	4.22
29	Bagalduba	5.00	3.50	2.67	3.00	—	3.54
30	Dudh	4.20	4.71	7.00	2.00	—	4.48

Though the average values of bifurcation ratios of 27 basins confirm the characteristics of natural stream system yet some variations within the orders of the basins are well marked. The maximum bifurcation ratios between the 5th and 6th orders (in the case of 6th order basins) are 11.00 (Koina) and 9.00 (Roro) caused by the exceptionally elongated shapes of these basins and between 4th and 5th orders (in the case of 5th order basins) maximum bifurcation ratios stand at 10.0 (Banki), 6.0 (Gara III), 5.00 (Kuchai) and 5.00 (Toyargara). These basins, except Gara III which is located over the central plateau, lie over the western highlands. In general, the rivers of the western highlands exhibit significant departures from general rule of natural stream system because varying nature of reliefs has guided the branching of drainage network whereas the rivers of the central plateau are characterized by consistent values of bifurcation ratios with few exceptions.

The general trend of bifurcation ratios conforms with the Giusti and Schneider's hypothesis (1965) that bifurcation ratios within a region decrease with the increasing order ($R_{b1} > R_{bk}$, K=successive increasing order) wherein 18 basins (Konkoa, Kongera, Torlo, Suru, Sankh, Kharsoti, Sapghara, Karanjia, Gara III, Sasagara, Sona, Bhangbang, Bagalduba, Tentarigara, Mahadeosalgara, Kumbigara, Bodikuchagara and Bamni) show consistent decrease in bifurcation ratios with increasing orders with only increase in two orders, whereas the remaining 12 basins show mixed trend. In fact, the deviation from the above hypothesis occurs in the basins of the western and northern highlands where geology and relief have affected the branching of streams. The only exception is Dudh river which shows increasing trend in bifurcation ratios though it drains over Dhonjori highlands.

The second postulation of Giusti and Schneider (1965, p. 2) that 'basins of equal order but variable areas tend to have the smallest bifurcation ratios in the smallest areas; the ratio increases with increasing areas up to certain size beyond which the bifurcation ratios tend to become constant, may not be applied to the 30 basins of S.E. Chota Nagpur region because it is not the area which decides the branching patterns of the streams but several other factors like lithological characteristics, slopes, relative reliefs, stage of basin development etc. play dominant role in the branching of streams. For example, Tentarigara is the smallest basin among the 6th order basins but its mean bifurcation ratio (3.74) is more than Konkoa (3.54) and Kharsoti (3.68) basins. Similarly, Kantajuria is the smallest basin of all of the 5th order basins but its mean bifurcation ratios of 3.84 is more than Bagalduba (3.54) and Sasagara (3.73) basins. The coefficient of correlation between mean bifurcation ratios and basin areas of 30 basins stands at +0.45 which is not very significant as only 20 per cent observation is explained. The poor correlation of mean bifurcation with absolute relief ($r=+0.27$), relative relief ($r=+0.11$), average slope ($r=+0.16$). drainage density ($r=+0.11$), stream frequency ($r=+0.21$), basin length ($r=+0.56$) clearly indicates its dependence over geological structure, lithological characteristics, climate, plant cover etc.

Law of stream number

The regression lines* (Figs. 8.2 and 8.3) and Table 8.3 denote that the Horton's law of stream number (1945) 'that the number of stream segments of successively lower orders in a given basin tends to form a geometric series beginning with the single segment of the highest order and increasing according to constant bifurcation ratio involving negative exponential function model'** holds good in the basins of Kumirta, Sapu, Bodikuchagara, Kumbigara, Ghagara, Kongera, Torlo, Sankh and Sapghara rivers of the 6th order wherein points nearly fall on the regression lines but remarkable departures are

* log y=log a-bx

y=stream number, x=stream order, a=constant,

b=regression coefficient.

** $N_u = R_b^{(k-u)}$

where N_u=number of stream segments of a given order

R_b=constant bifurcation ratio

K =highest order of the basin

Thus, $\sum N_u = \frac{R_b^k - 1}{R_b - 1}$

Or

$$\sum_{u=1}^{K} N_u = \frac{R_b^k - 1}{R_b - 1}$$

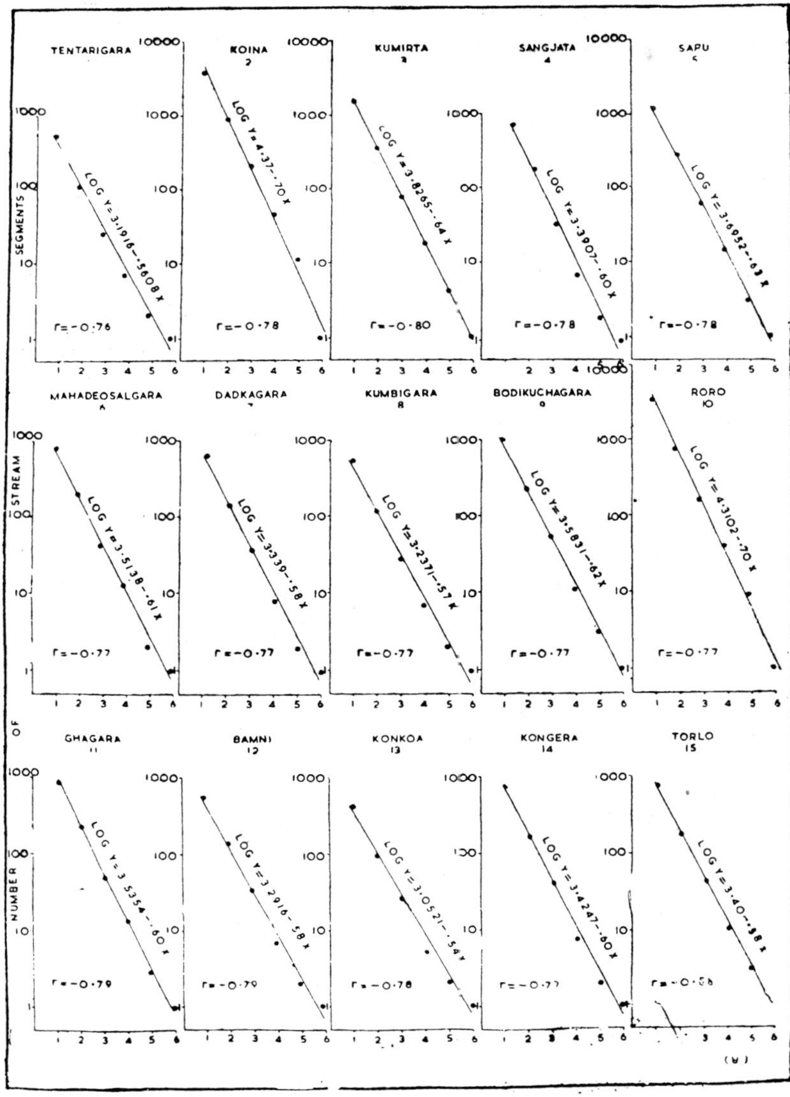

FIG. 8.2 : *The lawof stream number*

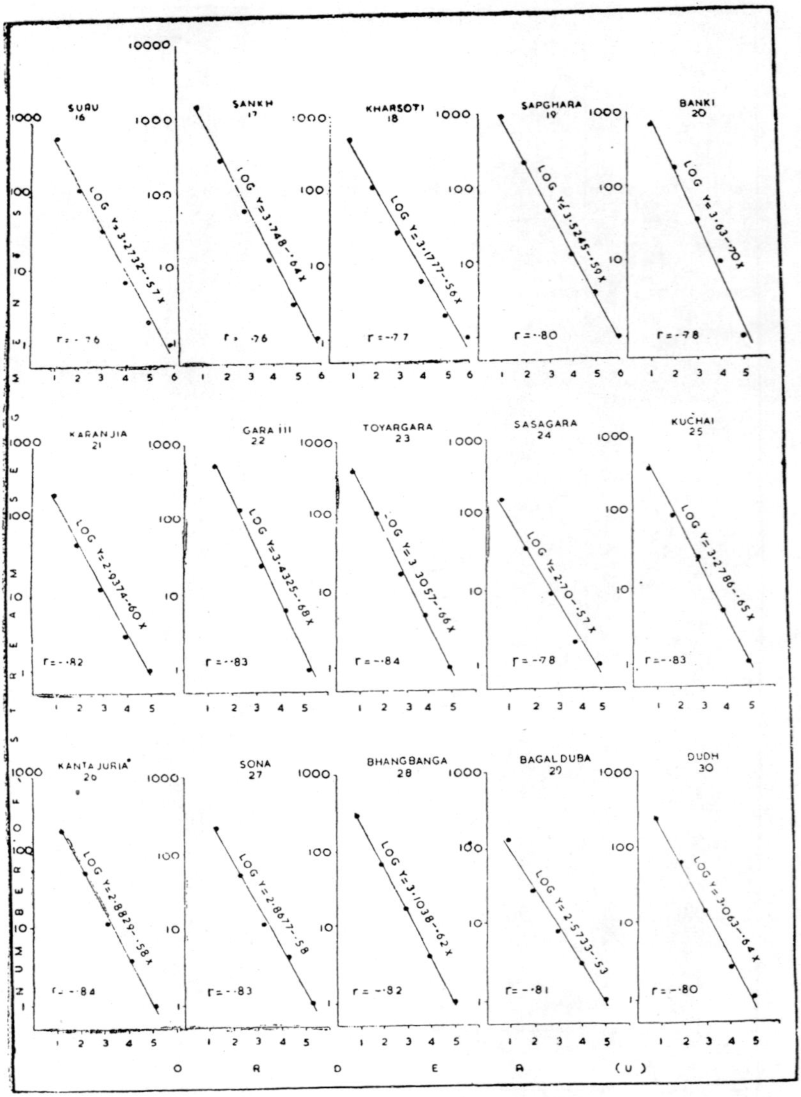

FIG. 8.3 : *The regression lines*

TABLE 8.3 : **Network relationships (topological) (coefficient of correlation)**

Sl. No.	Drainage Basins	Order (U) vs. Number of stream segments (N_u)	Order (U) vs. Cumulative mean length	Order (U) vs Mean basin area	Cumulative mean length vs. Mean basin area
1	Tentarigara	−0.76	+0.82	+0.80	+0.99
2	Koina	−0.78	+0.72	+0.71	+0.91
3	Kumirta	−0.80	+0.98	+0.93	+0.98
4	Sangjata	−0.78	+0.84	+0.88	+0.98
5	Sapu	−0.78	+0.78	+0.81	+0.99
6	Mahadeosalgara	−0.77	+0.93	+0.85	+0.99
7	Dadkagara	−0.77	+0.81	+0.86	+0.98
8	Kumbigara	−0.77	+0.86	+0.87	+0.98
9	Bodikuchagara	−0.77	+0.91	+0.82	+0.99
10	Roro	−0.77	+0.88	+0.87	+0.99
11	Ghagara	−0.79	+0.91	+0.81	+0.99
12	Bamni	−0.79	+0.96	+0.87	+0.99
13	Konkoa	−0.78	+0.96	+0.78	+0.94
14	Kongera	−0.77	+0.95	+0.86	+0.97
15	Torlo	−0.88	+0.90	+0.82	+0.99
16	Suru	−0.76	+0.91	+0.87	+0.95
17	Sankh	−0.76	+0.80	+0.82	+0.96
18	Kharsoti	−0.77	+0.81	+0.83	+0.96
19	Sapghara	−0.80	+0.88	+0.89	+0.99
20	Banki	−0.78	+0.79	+0.92	+0.999
21	Karanjia	−0.82	+0.81	+0.86	+0.91
22	Gara III	−0.83	+0.81	+0.89	+0.999
23	Toyargara	−0.84	+0.82	+0.90	+0.98
24	Sasagara	−0.78	+0.83	+0.81	+0.96
25	Kuchai	−0.83	+0.82	+0.90	+0.98
26	Kantajuria	−0.84	+0.82	+0.91	+0.93
27	Sona	−0.83	+0.89	+0.90	+0.99
28	Bhangbanga	−0.82	+0.94	+0.92	+0.998
29	Bagalduba	−0.81	+0.87	+0.81	+0.99
30	Dudh	−0.80	+0.84	+0.90	+0.98

All the values of 'r' are significant at 1% of probability level.

registered by the remaining 6th order basins (Tentarigara, Koina Sangjata, Mahadeosalgara, Dadkagara, Roro, Bamni,

Konkoa, Suru and Kharsoti basins) mainly in their higher orders because either they receive more tributaries or less streams in their lower reaches than required for ideal conditions. Koina and Roro basins represent the former conditions by receiving 11 and 9 tributaries in their 5th order respectively whereas remaining basins receive only 2 tributaries in their 5th orders. Such departures have also been noted by Strahler (1964), Leopold, Wolman and Miller (1964), Coates (1958), Singh and Srivastava (1974, 1976), Singh (1978) etc. Out of 11 basins of the 5th order, 10 basins completely correspond with the model of negative exponential function.

The correlation coefficients between the number of stream segments and orders (Table 8.3) range between—0.76 to—0.80 among the 6th order basins and between—0.78 to—0.84 among the 5th order basins, all of which are significant at 1 per cent of probability level.

Law of stream length

Generally, the 1st order stream segments have the shortest length but the mean length increases with the increasing order. All the 30 streams follow this postulation (Table 8.4) except some departures in certain orders of a few streams only. Suru and Kharsoti basins register departure in the 6th order where the mean length of the 6th order segment is less than the 5th order segment whereas Kumbigara registers decrease in the 5th order segment and Kantajuria and Karanjia indicate basins decrease in the 4th order segments.

The mean length ratios* of 30 drainage basins vary from 1.79 (Konkoa basin) to 4.74 (Koina basin) but a considerable variation is noted within the different orders of the basin (Table 8.5). It is apparent from Table 8.5 that length ratios neither increase nor decrease consistently with the orders to form a definite system. The only generalisation, which emerges from Table 8.5, is that the rivers, which drain over the peneplain surface of the central plateau region, have more or less

* $R_l = \dfrac{\bar{L}_u}{\bar{L}_{u-1}}$ when $\bar{L}_u = \dfrac{\Sigma L_u}{N_u}$

where $\bar{L}_u$=mean length of all stream segments of an order and N_u=number of stream segments

TABLE 8.4 : **Mean lengths ($\overline{L_u}$) of stream segments (kms)**

Sl. No.	Basins	$\overline{L_1}$	$\overline{L_2}$	$\overline{L_3}$	$\overline{L_4}$	$\overline{L_5}$	$\overline{L_6}$
1	Tentarigara	0.21	0.39	0.84	2.00	2.53	8.77
2	Koina	0.33	0.48	1.51	2.45	5.22	80.05
3	Kumirta	0.31	0.57	1.39	3.36	8.11	18.74
4	Sangjata	0.26	0.69	1.73	5.61	9.79	17.19
5	Sapu	0.23	0.43	1.02	2.89	3.86	12.39
6	Mahadeosalgara	0.24	0.40	0.89	2.07	4.14	9.25
7	Dadkagara	0.25	0.40	0.58	1.58	7.04	8.05
8	Kumbigara	0.21	0.30	0.58	2.11	5.04	9.33
9	Bodikuchagara	0.23	0.34	0.94	2.11	5.04	10.30
10	Rolo	0.23	0.37	1.29	2.30	5.85	37.65
11	Ghagara	0.24	0.39	1.01	2.65	5.04	8.61
12	Bamni	0.32	0.51	0.98	2.10	6.11	13.68
13	Konkoa	0.36	0.58	1.46	4.28	5.99	2.98
14	Kongera	0.51	1.13	2.59	6.74	6.19	12.20
15	Torlo	0.48	0.86	1.80	3.38	4.93	13.80
16	Suru	0.46	0.69	1.39	2.55	10.86	3.38
17	Sankh	0.19	0.38	2.10	5.58	11.13	13.27
18	Kharsoti	0.43	0.96	2.19	5.51	19.59	6.28
19	Sapghara	0.48	0.68	2.00	2.99	6.09	18.42
20	Banki	0.17	0.31	0.80	1.55	15.77	—
21	Karanjia	0.46	0.85	2.50	2.47	16.09	—
22	Gara III	0.57	1.09	3.19	3.78	36.04	—
23	Toyargara	0.35	0.54	1.40	2.25	17.54	—
24	Sasagara	0.41	0.86	1.34	2.57	16.73	—
25	Kuchai	0.48	0.89	2.14	5.21	31.21	—
26	Kantajuria	0.38	0.47	1.26	1.13	11.58	—
27	Sona	0.51	0.98	1.68	3.46	16.57	—
28	Bhangbanga	0.55	0.98	2.54	7.08	14.16	—
29	Bagalduba	0.41	0.77	1.07	1.88	8.53	—
30	Dudh	0.58	0.86	2.54	3.74	18.58	--

constant length ratios within the different orders whereas the rivers of the mountainous regions with scarp faces and much reliefs of the western highlands, Dalma lavas and Dhanjori highlands, present considerable variations in length ratios within the orders indicating their dependence on reliefs as the reliefs and watersheds are largely consumed by the lateral planation giving way to systematic growth of drainage network in orderly manner but the reliefs and resistant structure taboo the natural and systematic growth of drainage network.

TABLE 8.5 : **Length ratio (R_l)**

Sl. No.	Basins	$\overline{L_2}/\overline{L_1}$	$\overline{L_3}/\overline{L_2}$	$\overline{L_4}/\overline{L_3}$	$\overline{L_5}/\overline{L_4}$	$\overline{L_6}/\overline{L_5}$	$\overline{R_l}$
1	Tentarigara	1.86	2.15	2.38	1.27	3.47	2.23
2	Koina	1.45	3.15	1.62	2.13	15.34	4.74
3	Kumirta	1.84	2.44	2.42	2.41	2.31	2.28
4	Sangjata	1.65	2.42	3.73	1.08	1.77	2.13
5	Sapu	1.87	2.37	2.83	1.34	3.21	2.32
6	Mahadeosalgara	1.67	2.23	2.33	2.00	2.23	2.09
7	Dadkagara	1.60	1.54	2.72	4.46	1.14	2.27
8	Kumbigara	1.83	1.93	3.64	0.90	4.94	2.57
9	Bodikuchagara	1.48	2.76	2.24	2.39	2.04	2.18
10	Roro	1.61	3.49	1.78	2.54	6.44	3.17
11	Ghagara	1.63	2.59	2.62	1.90	1.71	2.09
12	Bamni	1.65	1.92	2.14	2.91	2.24	2.17
13	Konkoa	1.61	2.52	2.93	1.40	0.50	1.79
14	Kongera	2.22	2.29	2.60	0.92	1.97	2.00
15	Torlo	1.79	2.09	1.88	1.46	2.80	2.04
16	Suru	1.50	2.01	1.83	4.26	0.31	1.98
17	Sankh	2.00	5.53	2.66	1.99	1.92	2.82
18	Kharsoti	2.23	2.28	2.52	3.50	0.32	2.18
19	Sapghara	2.42	2.94	1.50	2.04	3.02	2.38
20	Banki	1.82	2.58	1.94	10.17	—	4.13
21	Karanjia	1.85	2.94	1.00	6.51	—	3.08
22	Gara III	1.91	2.93	1.18	9.53	—	3.89
23	Toyargara	1.54	2.59	1.61	7.80	—	3.39
24	Sasagara	2.10	1.56	1.92	6.51	—	3.02
25	Kuchai	1.85	2.40	2.43	5.99	—	3.17
26	Kantajuria	1.24	2.68	0.92	10.25	—	3.77
27	Sona	1.92	1.71	2.06	4.79	—	2.62
28	Bhangbanga	1.78	2.59	2.79	2.00	—	2.29
29	Bagalduba	1.88	1.39	1.76	4.54	—	2.39
30	Dudh	1.45	2.95	1.47	4.97	—	2.72

There is an inverse relationship between the total stream lengths of orders and orders of the basin wherein the total stream lengths (ΣL_u) continue to decrease from the 1st order to the successively higher orders. The general trend follows this rule but some deviations are noted in certain orders in a number of streams. The mean stream length ($\overline{L}_u$) and order present a direct relationship where mean stream length of stream segments increases with increase in orders. This law is again applicable to the 24 basins out of the 30 selected basins, the only exceptions are noted in Suru and Kharsoti (decrease between 5th and 6th order), Kantajuria and Karanjia (decrease from 3rd to 4th order) and Kumbigara and Kongera (decrease from 4th to 5th order) basins.

The regression lines* (Figs. 8.4 and 8.5), drawn to test the law of stream length of positive exponential function model of Horton (1945)** that 'the cumulative mean lengths of stream segments increase in geometrical progression with successive increase in stream orders with constant length ratio' indicate the fact that Tentarigara, Kumirta, Mahadeosalgara, Kumbigara, Bodikuchagara, Roro, Ghagara, Bamni, Konkoa, Kongera and Sapghara basins directly correspond to the positive exponential function model because nearly all the points come closer to the regression lines. Some pronounced deviations are well marked in Koina, Sangjata, Sapu, Dadkagara, Torlo, Suru, Sankh and Kharsoti basins where points depart from the regression lines. The trend of regression lines indicates the fact that the rivers of the highly dissected western highlands, Dalma lavas and Dhanjori highlands show deviations whereas all the rivers of the central plateau except Torlo river confirm the law of stream length. The 5th order basins also confirm the Hortonial model of stream length as Kuchai, Sona, Bhangbanga, Sasagara, Bagalduba, Karanjia and Dudh rivers do not exhibit any deviation from the above law as all the points nearly fall on the regression lines but pronounced variations are noted in Banki, Toyargara and Kantajuria basins

* $\log y = \log a + bx$

where y = cumulative mean length

x = stream order

b = regression coefficient

a = constant

** $\overline{L}_u = \overline{L}_1 R_l^{(u-1)}$

where $\overline{L_1}$ = mean length of the 1st order

R_l = constant length ratio

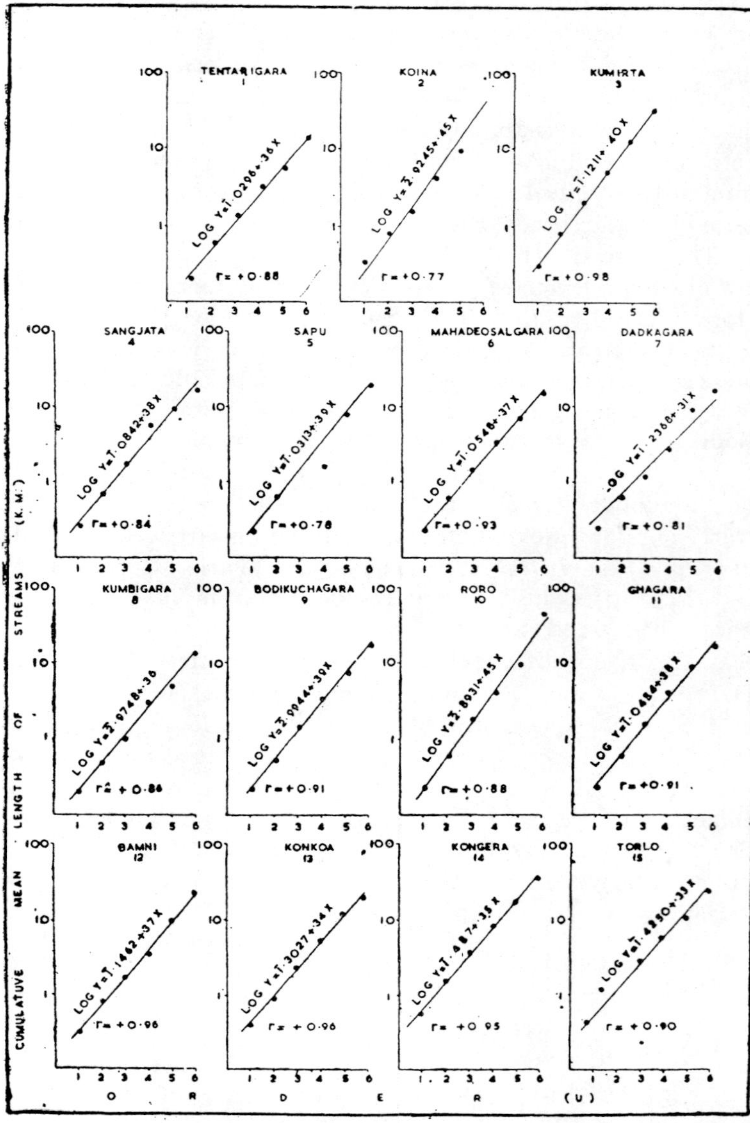

FIG. 8.4 : *The regression lines*

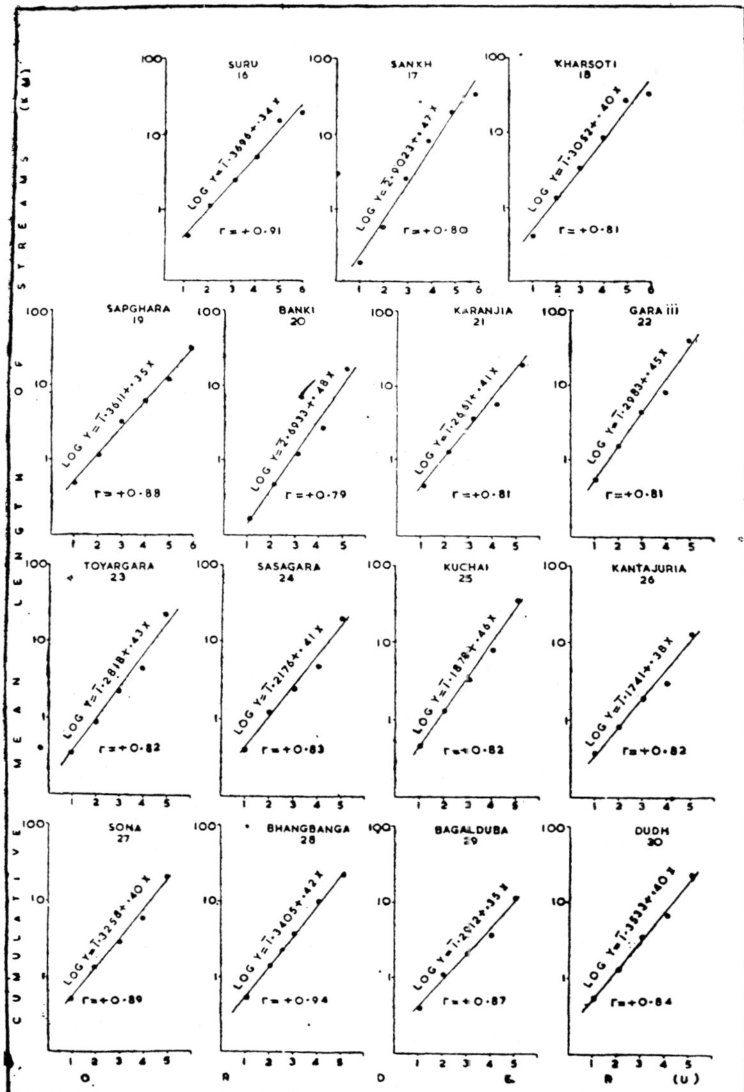

FIG. 8.5 : *The law of stream length of positive exponential function*

where points of the higher orders show departure from the regression lines. The coefficients of correlation between cumulative mean lengths and orders of 30 basins (Table 8.3) ranging between +0.72 to +0.98 (all significant at 1 per cent of probability level) validate the strong positive relationship between the aforesaid variables.

Law of basin area

The 1st order basins have the smallest mean basin areas and the successive higher orders show increase in the areas culminating in the largest area of the highest order of the trunk stream. It is apparent from Table 8.6 that basin areas are not affected either by lithological characteristics and geological formations or by the stage of the basin development as among the 6th order basins, the highest and the lowest positions in relation to the basin area are occupied by Koina (551.67 sq. kms.) and Tentarigara (48.07 sq. kms.) basins both of which are located in the western highlands. Kongera occupies the second position in the scale of areal extent (449.98 sq. kms.) which is located in the central plateau having granite-gneissic formation and is placed in the penultimate stage of basindev elopment. Similarly, 5th order basins also do not follow any systematic rule in relation to basin area. It seems that the basin areas are highly affected by the nature of the terrain. In the western highlands, only those basins have larger basin areas which are bounded by the parallel hill ranges on both the sides. This observation becomes true in the case of Koina and Roro basins which have developed elongated basins due to the presence of parallel hill ranges. The basins of the central plateau, except Kongera, are guided by the dolerite dykes which have given elongated shapes to the basins of this region and have helped them to occupy larger basin areas. The basins of the Dalma lavas and Dhanjori highlands have developed larger areas by eroding the conglomerates, phyllites and epidiorites of these regions.

The regression lines (Figs. 8.6 and 8.7) drawn to test the validity of positive exponential function model* of law of basin

* $\bar{A}_u = \bar{A}_1 \, R_a^{(u-1)}$

where $\bar{A}_1$ = mean area of the 1st order

R_a = constant area ratio

TABLE 8.6 : **Basin area (A_u) (in sq. kms.)**

Sl. No.	Basins	ORDER (U)					
		A_{u1}	A_{u2}	A_{u3}	A_{u4}	A_{u5}	A_{u6}
1	Tentarigara	28.95	37.21	42.12	44.84	46.66	48.07
2	Koina	316.08	412.02	470.21	521.42	544.60	551.67
3	Kumirta	154.60	226.21	264.12	291.45	302.89	311.19
4	Sangjata	44.52	57.04	63.84	68.10	71.20	74.82
5	Sapu	74.40	94.78	108.54	120.46	126.42	131.78
6	Mahadeosalgara	53.76	70.12	78.58	84.56	88.28	92.77
7	Dadkagara	26.76	34.85	40.24	44.94	46.82	49.99
8	Kumbigara	29.70	37.12	42.50	44.87	47.24	49.16
9	Bodikuchagara	44.36	58.20	65.40	71.41	76.42	78.99
10	Roro	147.76	221.06	265.48	301.54	329.12	340.71
11	Ghagara	61.39	80.27	92.58	101.12	106.10	109.76
12	Bamni	53.10	68.21	75.85	81.08	86.52	89.41
13	Konkoa	57.68	74.28	85.12	92.43	100.28	106.37
14	Kongera	165.48	265.20	330.10	370.25	408.58	449.98
15	Torlo	142.50	192.50	212.85	228.00	240.00	252.91
16	Suru	56.79	69.24	84.56	96.25	105.10	112.33
17	Sankh	191.92	255.85	298.91	325.45	342.56	354.52
18	Kharsoti	112.14	180.21	205.54	219.64	230.10	237.56
19	Sapghara	132.30	185.64	213.12	228.21	239.42	245.67
20	Banki	36.30	46.05	50.12	52.85	54.05	—
21	Karanjia	48.60	60.12	70.58	82.82	89.67	—
22	Gara III	163.37	210.59	241.21	261.57	294.30	—
23	Toyargara	33.92	45.51	55.12	62.25	67.99	—
24	Sasagara	33.39	44.58	52.45	58.48	62.65	—
25	Kuchai	76.95	98.89	121.12	141.57	154.62	—
26	Kantajuria	18.45	23.37	26.51	29.98	32.01	—
27	Sona	45.15	62.12	73.23	84.25	94.43	—
28	Bhangbanga	78.75	110.25	130.21	149.37	163.69	—
29	Bagalduba	19.60	25.32	29.25	32.84	35.30	—
30	Dudh	55.40	78.12	96.54	101.11	110.18	—

area of Strahler (1969, p. 488) as the 'mean basin areas of successive increasing orders tend to form a geometric series beginning with the mean area of the 1st order basin and increasing according to constant area ratio', indicate the fact that out of 19, 6th order basins, only 6 basins (Sapu, Roro, Kumirta, Torlo, Sapghara and Mahadeosalgara) closely conform with

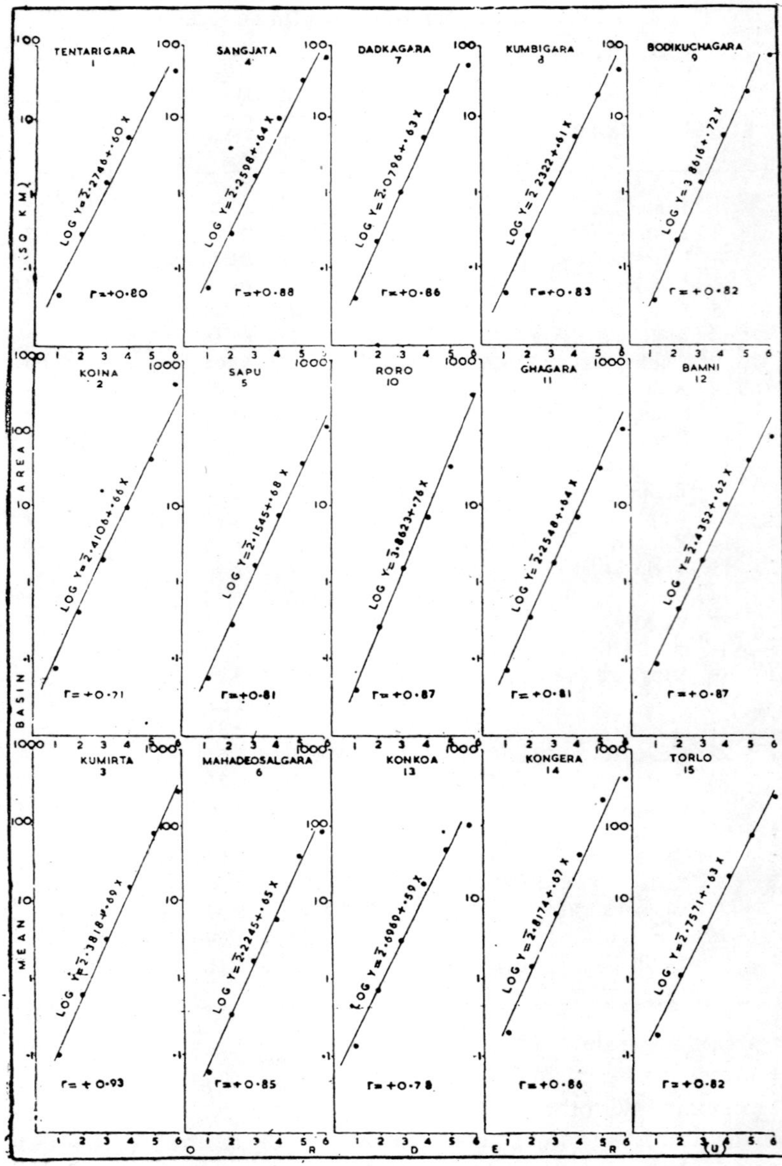

FIG. 8.6 : *The regression lines drawn to test the validity of positive exponential function*

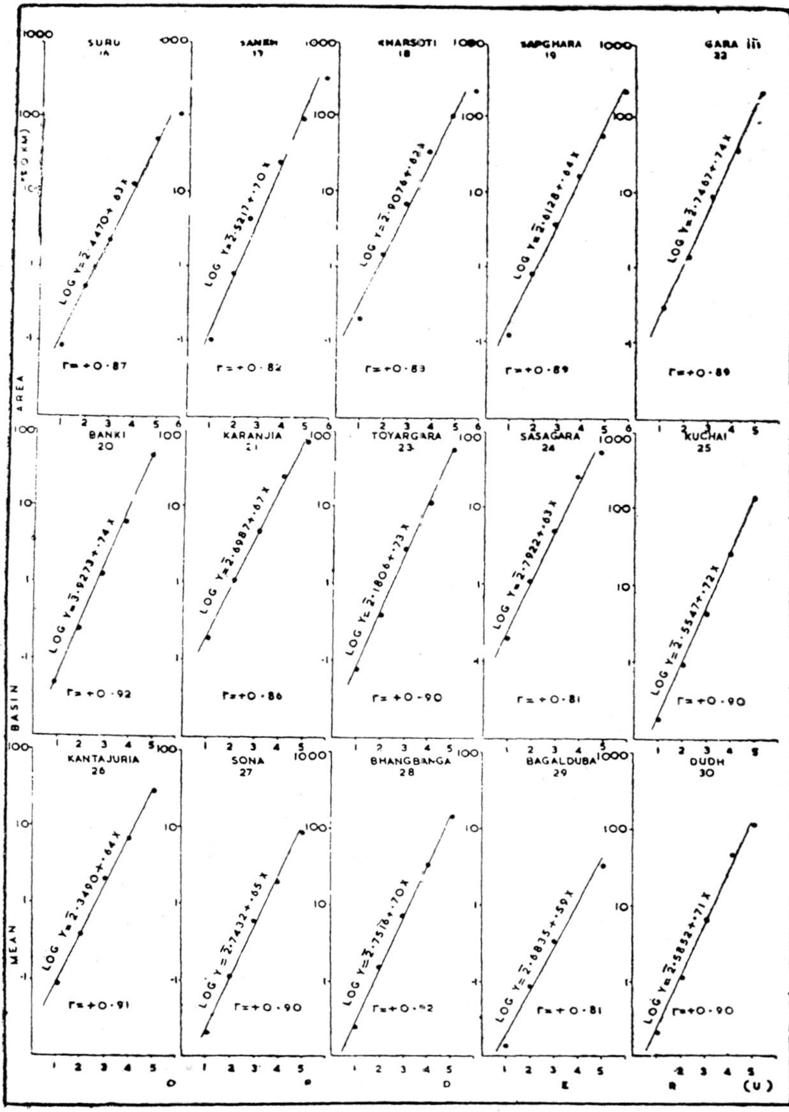

FIG. 8.7 : *The regression lines*

the above law whereas the remaining 13 basins present remarkable departure from the said law. Such departure was even noted by Strahler himself (1969, p. 488) during his study of Allegheny and Fern Canyon rivers. Singh and Srivastava (1974 and 1976) also observed such deviations during their studies of tributary basins of the Belan river located over Vindhyan sandstones. Singh (1978) again found such deviations from the above law during his study of 23 drainage basins of the Ranchi plateau having more or less similar geological structure as that of the S.E. Chota Nagpur region located just to the north-west of the latter. The result is somewhat different in the cases of the 5th order basins wherein 9 basins (Gara III, Banki, Karanjia, Toyargara, Kuchai, Kantajuria, Sona, Bhangbanga and Dudh basins) out of 11, 5th order basins, correspond to the above law. The departures of the points from the regression lines are due to the variations in area ratios within the orders of the basins whereas the law requires a constant area ratio which is not possible in nature but the generalization confirms the above law of linear relationship between mean basin areas and orders as the coefficients of correlation between these two variables of 30 basins range between +0.71 and +0.93 (Table 8.3) all of which are significant at 1 per cent of probability level.

Law of allometric growth

Recently Woldenberg (1966), Strahler (1969) and other geomorphologists have introduced the law of allometric growth of organs of plants and animals in the fluvial system based on the analogy that the measurable properties of a drainage basin viz. stream length and basin area can be compared with the organs of growing plant or animal and their increase is correlated with the law of allometric growth. The successive increasing orders, with the advancement of fluvial cycle, indicate increasing time where appearance of new 1st order finger-tip stream segments, with accelerated rate of erosion during youthful stage, increases the order of the older segments. Under ideal conditions of uniform geological structure and climate, the rate of increase (growth) becomes orderly and follows the law of allometric growth indicating a positive power function model of mean basin area and cumulative mean lengths.

It is apparent from Figs. 8.8, 8.9, 8.10 that Tentarigara, Kumirta, Sangjata, Sapu, Mahadeosalgara, Dadkagara, Bodikuchagara, Roro, Ghagara, Konkoa, Torlo, Sankh, Kharsoti and Sapghara rivers of the 6th order closely conform to the law of allometric growth because nearly all the points fall nearer to the regression lines.* Some departures are noted in Koina, Kumbigara, Kongera and Suru basins which may be treated as chance occurrences. Similarly, 8 out of 11, 5th order basins (Banki, Karanjia, Gora III, Kantajuria, Sona, Bhangbanga, Bagalduba and Toyargara) closely correspond to the above law. Some remarkable variations of points from the regression lines are noted in Sasagara, Kuchai and Dudh basins where the points of the highest order significantly depart from the regression lines. The coefficients of correlation (Table 8.3) between mean basin areas and cumulative mean lengths of 30 basins ranging between +0.91 and +0.999 (significant at 1 per cent of probability level) also confirm the validity of above law.

This law of basin morphometry indicating proportionate growth between mean basin areas and cumulative mean lengths with the march of time clearly interprets the genetic aspect of basin evolution wherein it is axiomatic to deduce that during a normal functioning of stream-work, the back wearing of divides and interstream areas through the process of lateral erosion results into gradual increase of basin areas whereupon the stream lengths are increased due to headward erosion.

The following observations may be made on the basis of network analysis of 30 drainage basins of the S.E. Chota Nagpur region :

(i) The laws of stream number (negative exponential function model), stream lengths (positive exponential function model) and basin areas (positive exponential function model) and allometric growth (positive power function model) are more closely applicable to the

* $\log y = \log a + b \log x$
  y = mean basin area, x = cumulative mean length

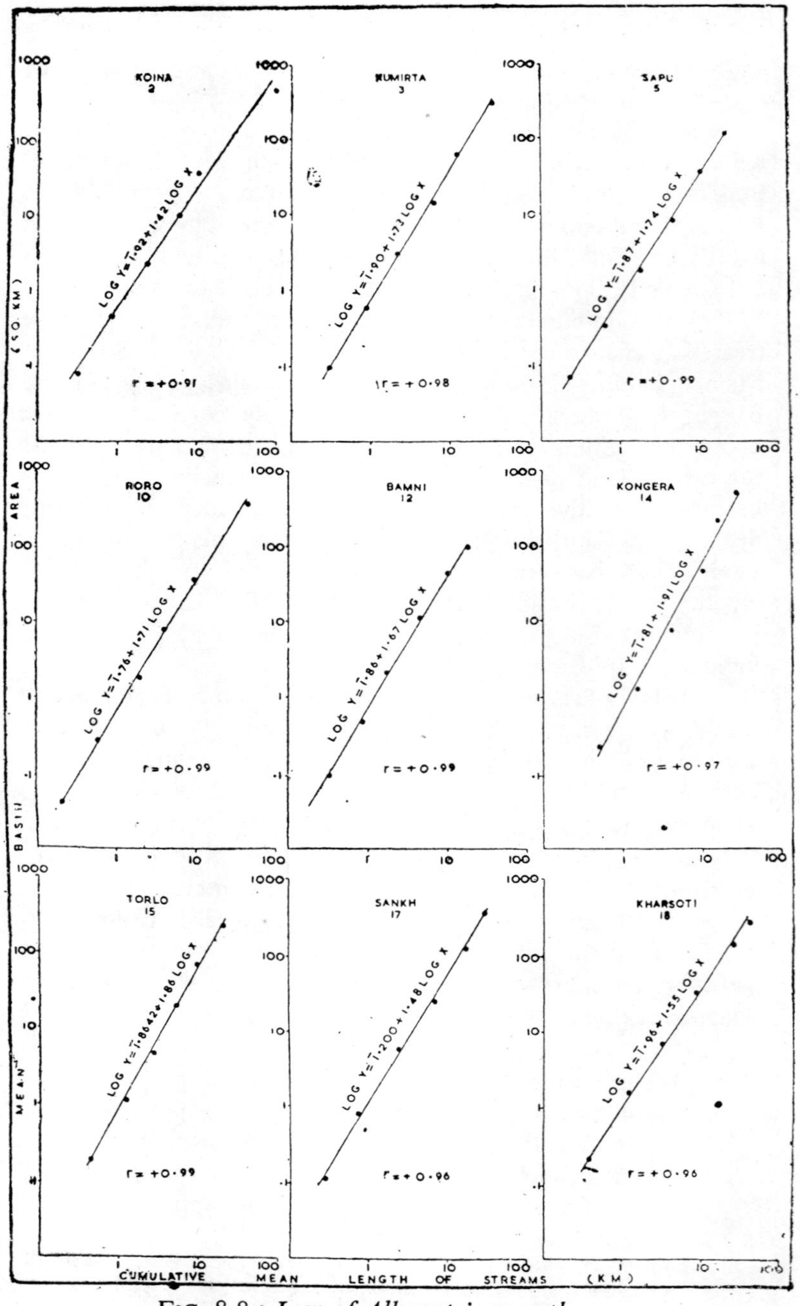

FIG. 8.8 : *Law of Allometric growth*

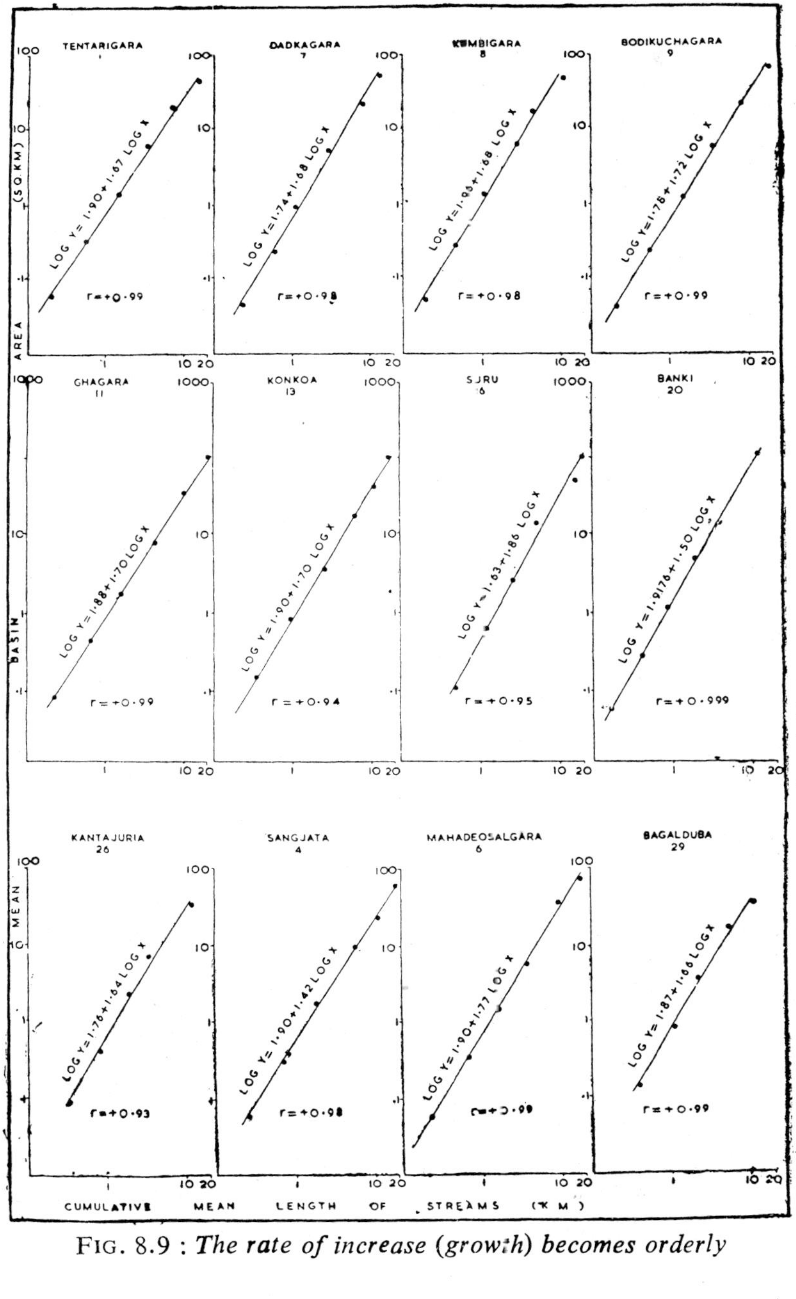

FIG. 8.9 : *The rate of increase (growth) becomes orderly*

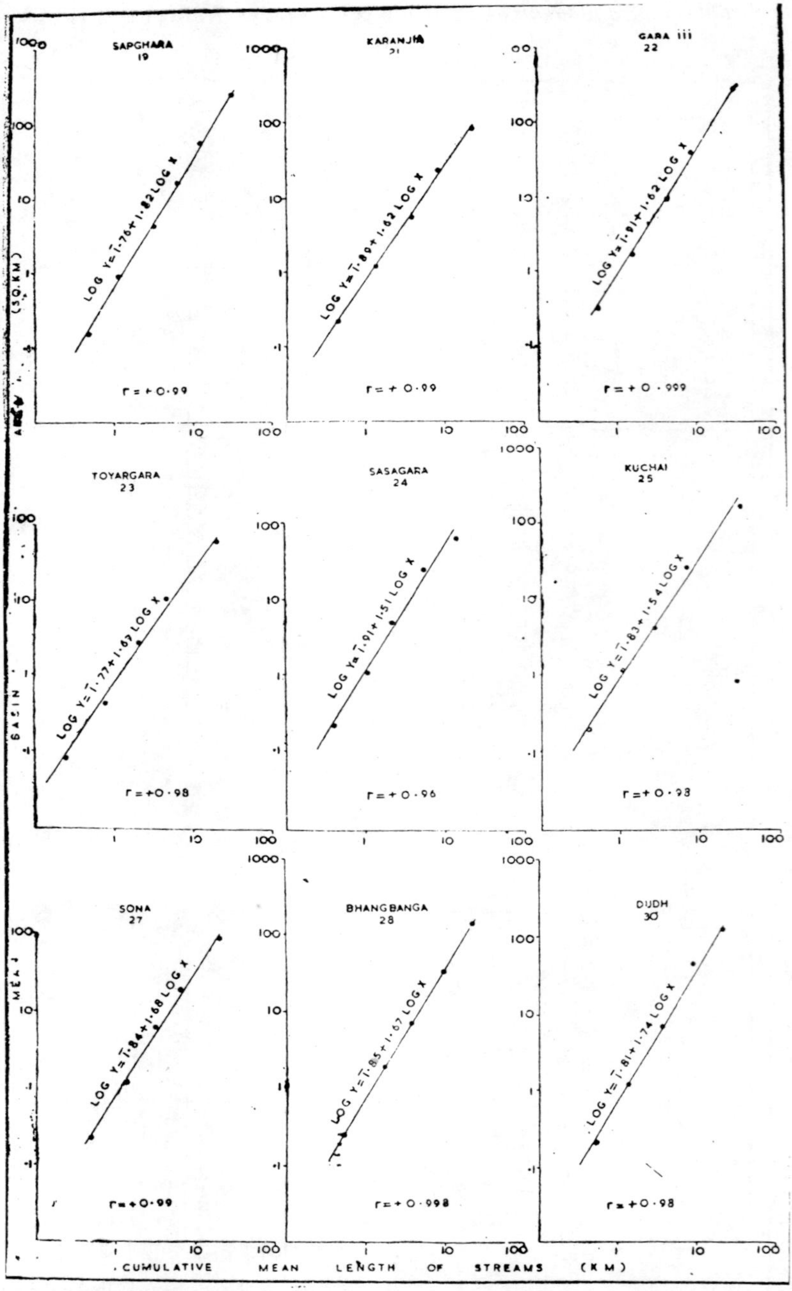

FIG. 8.10 : *The law of Allometric growth indicating a positive power function*

network of the basins more advanced in the stages of basin development than to the basins of transient stage (youthful stage) as is evident from the regression lines of the basins of the different physiographic regions.

(ii) The lower order basins having homogeneity in geological structure closely conform to the above laws more than the basins of the higher order.

(iii) The more the heterogeneity in the terrain characteristics, the more is deviation in network laws.

Law of basin perimeter, basin length and basin area

Basin area, basin perimeter and channel length are the significant morphometric variables which determine the shape, size and genetic aspect of drainage basins. Basin perimeter can be directly correlated with the square roots of basin area and increase or decrease in the former indicates increase or decrease respectively in the latter. The coefficient of correlation between these two variables of 30 drainage basins standing at +0.99 is very strong positive correlation significant at 1 per cent of probability level. Similarly coefficients of correlation between basin perimeters and channel lengths and between square roots of basin areas and channel lengths being +0.89 and +0.98 respectively are again significant at 1 per cent of confidence level. The significant relationships among the aforesaid variables clearly speak the story of genetic aspect of basin development. The first order segments of the basin notch down the watersheds from which they take their sources through headward erosion. This process helps in the proportionate increase of basin area and the back-wearing of watersheds favours the increase of stream length and basin area alike. The back-wearing of divides simultaneously increases the basin perimeter. This law works in the region of homogeneous geological structure, climate and plant cover but significant variation in relief characteristics upsets the smooth functioning of basin development and introduces departures from the above law as is the case of Kongera and Konkoa basins.

Length of overland flow (L_g)

The length of overland flow, considered as a dominant hydrologic and morphometric factor, is 'the mean horizontal length of flow-path from the divide to the stream in a first order basin and is a measure of stream spacing and degree of dissection and is approximately one-half the reciprocal of the drainage density, (Chorley, 1969). It is apparent from Table 8.7 that the values of L_g vary from 0.07 (Dadkagara and Kumbigara basins) to 0.16 (Kongera and Bhangbanga basins). The spatial analysis of L_g (Table 8.7) shows a marked variation in the values of L_g of 30 basins draining over different physiographic regions. The basins of the western highlands are marked with the minimum Lg whereas the basins of the central plateau are characterized by maximum L_g. The basins of the western part of the Dalma highlands have comparatively lower values of L_g (Suru =0.09 and Kantajuria =0.08) than the basins of the eastern part of the same region (Kharsoti =0.14, Sapghara =0.13 and Bagalduba =0.12) while Sankh (0.14) and Dudh (0.13) basins of the Dhanjori highlands have higher values of length of overland flow.

TABLE 8.7 : **Lengths of overland flow (in mile)**

Sl. No.	Basins	L_g	Sl. No.	Basins	L_g
1	Tentarigara	0.08	16	Suru	0.09
2	Koina	0.09	17	Sankh	0.14
3	Kumirta	0.12	18	Kharsoti	0.14
4	Sangjata	0.08	19	Sapghara	0.13
5	Sapu	0.08	20	Banki	0.08
6	Mahadeosalgara	0.08	21	Karanjia	0.14
7	Dadkagara	0.07	22	Gara III	0.15
8	Kumbigara	0.07	23	Toyargara	0.09
9	Bodikuchagara	0.08	24	Sasagara	0.14
10	Roro	0.09	25	Kuchai	0.13
11	Ghagara	0.08	26	Kantajuria	0.08
12	Bamni	0.09	27	Sona	0.14
13	Konkoa	0.12	28	Bhangbanga	0.16
14	Kongera	0.16	29	Bagalduba	0.12
15	Torlo	0.14	30	Dudh	0.13

The spatial analysis of distribution of L_g reveals the fact that the basins more advanced in the cyclic stage are characterized by higher values of L_g as is the case of the basins of the central plateau but the basins of the western highlands are in late youth to early mature stages of the basin development and are characterized by the lowest values of L_g but the basins of the central plateau clearly exhibit the symptoms of the senile stage and have the highest values of L_g. This observation contradicts the finding of Coates (1958) who related the values of L_g with the stage of basin development by stating that 'other factors being constant, areas more advanced into maturity appear to contain smaller overland flow than youthful areas'. This postulation of Coates is not tenable because the L_g is directly negatively correlated with drainage density; the higher the drainage density, the smaller the L_g and vice-versa and 'since a drainage basin on an average develops maximum stream segments in its late youth and early mature stages, hence, minimum length of overland flow must be found (Singh 1978). Singh (1978) studied the length of overland flow of 23 drainage basins of the Ranchi plateau located just to the N.W. of the present study region and found that the basins of the western highland (Pat region) of the Ranchi plateau, which present the features of the late youth to early mature stages, were characterized by the shorter length of overland flow than the basins of the peneplain stage of the central Ranchi plateau.

The length of overland is strongly negatively correlated with drainage density ($r = -0.89$), strongly positively correlated with stream frequency ($r = +0.87$) and with drainage texture ($r = +0.90$), negatively correlated with average slope ($r = -0.75$) and with dissection index ($r = -0.38$). The negative relationship between average slope and L_g indicates that the steeper the slope angles, the smaller the lengths of overland flow and vice versa.

In fact, the L_g is largely controlled and affected by geological formation, lithological and pedological characteristics, rainfall intensity, infiltration rate, plant cover etc. Generally dense plant cover is antagonistic to overland flow as some part of falling rain is subjected to dissipation in the air and the intensity of remaining rainwater is reduced due to breaking of rain drops because of obstruction offered by the plant cover and, thus, the packing of soil to form an impermeable crust is discouraged. The result is more infiltration of water and marked reduction in overland flow. Further, some part of rainfall is trapped by the plant canopy causing 'areal streams' wherein water reaches the ground through the branches and stems of trees and increases infiltration in the illuviation zone of the soil and, thus, the overland flow is reduced. This observation is validated in the cases of Koina, Tentarigara, Sapu, Sangjata, Mahadeosalgara, Kumbigara, Bodikuchagara, Roro, Banki, Dadkagara and Bamni basins draining over the western highlands and Suru and Kantajuria located in Dalma highlands having thick plant covers of moist and dry deciduous forests wherein the length of overland flow varies from 0.07 mile to 0.09 mile. On the other hand, over a barren surface generally free from plant cover the rain drops fall without any obstruction with the result they are neither impeded nor split rather they strike the surface directly due to which 'fine material is thrown into suspension by the impact and is redeposited as an almost impermeable surface skin which can lower infiltration by as much as ten times, (Kirkby, 1969, p. 227). Secondly, 'when the rainfall reaches a certain intensity, it exceeds the rate of ground absorption and overland flow occurs while the subsoil remains dry' (Tricart and Cailleux, 1972, p. 75). This rule works over the central plateau of bare granite-gneissic structure having scant plant cover mainly shrubs and bushes which have decreased the infiltration and have increased the overland flow as may be seen in the basins of Kongera, Konkoa, Gara III, Karanjia, Sasagara, Kuchai, Torlo, Sona and Bhangbanga rivers wherein L_g varies from 0.12 mile to 0.16 mile.

Shapes of the open link

The shapes of the open link in terms of geometric structure of drainage lines involves the calculation of departure of observed path (O_l) from the expected path (E_l) of a river from its source to its mouth. It is supposed that the expected path of the river will be a straight line but practically the straight line path of a river is never possible because it is affected by a number of causative factors viz. geological and hydrological controls, dip angles, slopes, absolute reliefs, relative reliefs, stage of valley development etc. which force the drainage line to deviate from its straight line expected path. The river courses are classified into three categories viz. (i) straight river course (SSI=1.00), (ii) sinuous course (SSI*=1.0 to 1.5) and meandering courses (SSI=>1.5).

Standard sinuosity

It is apparent from Table 8.8 that no river of the study region has meandering course because the SSI of all the 30 drainage basins range between 1.02 and 1.28 indicating sinuous course of all the basins. Though the rivers of the central plateau are in senile stage of the basin development but the granitic-gneissic structure of the plateau has presented a great obstacle in the development of flood plains and meanders by the small basins but they have developed at least sinuous courses wherein Kongera, Gara III and Bhangbanga register highest value of SSI (1.28). On the other hand, the basins of the western highlands viz. Tentarigara (1.08), Dadkagara (1.02), Kumbigara (1.07), Bodikuchagara (1.08), Roro (1.07), Ghagara (1.10) and Banki (1.04) have low value of SSI which is indicative of late youth to early mature stages of basin development. The rivers of the western highlands were rejuvenated due to Tertiary upliftments which is reflected in the low values of SSI of this region. Though Kumirta river flows over the western highlands yet it registers comparatively high value of SSI (1.19) which is due to the flattish surface of terrain of its basin. Bamni river emerging from the western highlands registers SSI of 1.1[illegible] which is the result of its lower reaches being located over flat surface of the central plateau. Dudh (1.17) and Sankh (1.19) basins of the Dhanjori highlands have SSI values more than for the basins of the western highlands but less than for the basins of the central

*SSI = Standard Sinuosity Index

$$= \frac{CL}{VL}$$

where CL = Channel length

VL = Valley length

TABLE 8.8 : **Sinuosity indices**

Sl.	Basins	CI	VI	HSI (%)	TSI (%)	SSI
1	Tentarigara	1.50	1.39	22.00	78.00	1.08
2	Koina	2.78	2.55	12.93	87.07	1.09
3	Kumirta	2.23	1.87	29.27	70.72	1.19
4	Sangjata	1.50	1.36	28.00	72.00	1.10
5	Sapu	1.65	1.57	12.31	87.69	1.05
6	Mahadeosalgara	1.49	1.46	6.12	93.88	1.02
7	Dadkagara	1.77	1.74	3.90	96.10	1.02
8	Kumbigara	1.43	1.34	20.93	79.07	1.07
9	Bodikuchagara	1.84	1.69	26.60	73.40	1.08
10	Roro	2.23	2.08	12.20	87.80	1.07
11	Ghagara	1.49	1.36	26.53	73.47	1.10
12	Bamni	2.40	2.16	17.14	82.86	1.11
13	Konkoa	1.96	1.56	41.60	58.40	1.26
14	Kongara	1.67	1.30	55.22	44.78	1.28
15	Torlo	1.70	1.35	50.00	50.00	1.26
16	Suru	1.98	1.89	9.19	90.81	1.05
17	Sankh	1.89	1.59	33.71	66.29	1.19
18	Kharsoti	1.61	1.35	42.63	57.37	1.19
19	Sapghara	1.68	1.48	29.41	70.59	1.14
20	Banki	1.87	1.79	9.08	90.92	1 04
21	Karanjia	1.77	1.60	27.07	77.93	1.11
22	Gara III	1.73	1.35	52.05	47.95	1.28
23	Toyargara	1.93	1.64	31.19	68.81	1.18
24	Sasagara	1.68	1.41	39.71	60.29	1.19
25	Kuchai	1.73	1.55	24.66	75.34	1.12
26	Kantajuria	1.74	1.65	12.16	87.84	1.05
27	Sona	1.92	1.52	43.48	56.52	1.26
28	Bhangbanga	1.93	1.51	45.17	54.83	1.28
29	Bagalduba	1.41	1.22	46.34	53.66	1.16
30	Dudh	1.70	1.45	35.71	64.29	1.17

plateau because of their late mature stage of basin development. The rivers of the Dalma highlands show mixed trends as Suru and Kantajuria basins located over the escarpment zone have SSI of 1.05 each while Kharsoti (1.19), Sapghara (1.14) and Bagalduba (1.16) basins spread over eastern comparatively level side of the Dalma highlands have slightly higher SSI values.

Hydraulic and topographic sinuosity

The hydraulic and topographic sinuosity indices (HSI and TSI)* are significant morphometric tools which help in determining the stages of basin development as well as in detecting the controlling factors of river sinuosity. It is axiomatic to deduce that the rivers have high percentages of TSI in their early stages of basin development which is replaced by high percentage of HSI in penultimate stage. Table 8.8 indicates the fact that the rivers of the central plateau register relatively high percentage of HSI which ranges between 27.07 to 55.22. These values, though not significant, reveal the fact that the courses of the basins of the central plateau are controlled by hydrological factors but topographic controls still persist due to the presence of resistant dolerite dykes in the form of 'dungris'. The sinuosity of the rivers of the western highlands is controlled by topographic factors as the TSI of different basins range between 70.73 and 96.10 per cent. The rivers of the Dhanjori and Dalma highlands exhibit combined effects of topographic and hydrological factors. Suru (90.81), Kantajuria (87.87) and Sapghara (70.59) present dominance of topographic controls whereas Kharsoti (TSI=57.37), Bagalduba (TSI=53.66), Sankh (TSI=66.29) and Dudh (TSI=64.29) basins though indicate the significance of topographic factors but the effects of hydraulic controls in these rivers cannot be ignored.

Meander properties

Though Muller (1968) has explained the controls of sinuosity of the rivers in a practical term in his model but it suffers

*HSI= % equivalent of $\frac{CI - VI}{CI - 1}$

TSI= % equivalent of $\frac{VI - 1}{CI - 1}$

where, $CI = \frac{CL}{Air\ L}$

$VI = \frac{VL}{Air\ L}$

where CI= Channel Index, VI=Valley Index

CL= Channel Length, VL=Valley Length

Air L= Shortest Straight Length

from the difficulty of arbitrary definition of meandering course of a river as pointed out by Brice (1964) who has presented his models* for the calculation and derivation of different components of the meandering course of the river. The high degree of symmetry of meander length, meander height and form ratios is the characteristics of meandering properties which are applied 'to a sinuous stream that has, from place to place along its course, one or a series of symmetrical arcs, the length of which is related to the width of the stream' (Brice, 1964, p. 26).

It is apparent from Table 8.9 that the rivers of the central plateau (Kongera, Konkoa, Bhangbanga, Sona, Sasagara, Gara III, Torlo, Karanjia and Toyargara) have high values of symmetry of meander length (40.15% to 59. 63%), meander height (41.23% to 59.62%) and symmetry of form ratio (41.26% to 58.12%). These values indicate the penultimate stage of valley development wherein the rivers develop their meandering course. Generally, the rivers, which have these values above 60 per cent, are termed as meandering rivers but the case of central plateau is somewhat different as the plateau is in peneplain stage yet the granitic and gneissic structures have restricted the development of full size of meanders. The rivers of the western highlands, Dalma and Dhanjori highlands present reverse conditions by having low percentages of symmetry of meander length (12.12% to 42.14%), meander height (22.16% to 48.24%) and form ratio (22.24% to 43.25%). These

* (i) Symmetry of (Arc) meander length $= 100 - \frac{100 \text{ (mean deviation of arc length)}}{\text{mean arc length}} \%$

(ii) Symmetry of (Arc) meander height $= 100 - \frac{100 \text{ (mean deviation of arc height)}}{\text{mean arc height}} \%$

(iii) Symmetry of form ratio $= 100 - \frac{100 \text{ (mean deviation of form ratio)}}{\text{mean form ratio}} \%$

(iv) Repetition of symmetry $= \frac{Na}{\Sigma Na} \times 100$

where Na = % of number of arcs whose length does not deviate more than 25% from the average length of meanders.

ΣNa = total number of meanders

TABLE 8.9 : **Meander property**

Sl. No.	Basins	Symmetry of meander length %	Symmetry of meander height %	Symmetry of form ratio %	Repetition of symmetry %
1	Tentarigara	30.25	31.16	35.54	18.23
2	Koina	42.14	48.24	37.24	21.54
3	Kumirta	30.14	29.21	26.24	11.14
4	Sangjata	32.15	31.17	30.24	20.24
5	Sapu	38.14	35.15	37.24	21.54
6	Mahadeosalgara	35.15	34.14	40.00	25.00
7	Dadkagara	36.16	35 26	35.14	20.38
8	Kumbigara	30.16	28.25	32.16	25.00
9	Bodikuchagara	26.12	25 23	26.24	16.14
10	Roro	35.16	35.00	36.21	20.14
11	Ghagara	34.25	31.16	36.25	32.24
12	Bamni	27.19	32.16	22.24	19.16
13	Konkoa	54.16	55.15	56.12	30.00
14	Kongera	52.16	47.60	51.41	50.12
15	Torlo	55.60	57.50	58.12	42.16
16	Suru	30.61	36.25	32.14	15.00
17	Sankh	32.25	39.14	31.16	12.00
18	Kharsoti	30.19	35.13	36.14	19.14
19	Sapghara	32.14	30.14	36.25	19.14
20	Banki	30.14	35.26	34.23	18.25
21	Karanjia	40.15	48.26	41.15	33.16
22	Gara III	59.63	50.21	54.20	50.24
23	Toyargara	46.15	41.23	45.25	38.15
24	Sasagara	47.16	42.21	41.26	30.00
25	Kuchai	40.15	30.14	46.23	22.24
26	Kantajuria	26.12	30.25	31.23	31.00
27	Sona	55.23	59.62	51.21	44.00
28	Bhangbanga	51.21	56.25	53.21	41.20
29	Bagalduba	38.23	34.21	36.33	21.14
30	Dudh	39.16	32.15	43.25	24.00

values indicate that terrain and lithological characteristics of rocks have restricted these rivers to develop meanders of expected size. Repetition of symmetry is of vital significance in deciding the meandering properties of the streams wherein the higher percentage of repetition of symmetry is closely related to the meandering course whereas the low percentage is indicative of straight or sinuous stream course. Table 8.9 repeats the same story as given in the foregoing paragraphs because the rivers of the central plateau have higher percentages of repetition of symmetry (30.00% to 50.24%) whereas the rivers of the other regions have low percentages (11.14% to 32.24%).

Thus, it may be concluded that the rivers of the central plateau are in the penultimate stage of basin development whereas the rivers of the other regions are in late youth to premature stage. It may be forwarded that the shape of open link of stream, say longitudinal course, largely depends on the geological structure and terrain characteristics as several hill ranges and hard and resistant geological formations have forced several rivers not to assume expected properties of meandering course even in their senile stage of basin development and, thus, it is futile to relate the meandering properties and sinuosity of the rivers with the stages of basin development.

Geometric structure of angular properties of open links

Geometry of angular property of drainage network is related to the junction angle of two stream segments which is defined as 'the angle, projected to the horizontal, between average flow-directions, determined by the ends of the stream segments extending from the junction point to an upstream point at a distance equal to 0.2 times the average length of the second order streams' (Lubowe, 1964). In the present study the entrance angles of the first order streams have been measured with the help of the horizontal inclination of two lines showing average flow-direction of the two tributaries of the 1st order. The same procedure has been applied for the calculation of the entrance angles of second, third, fourth and fifth orders of all the basins and the average entrance angles of all the orders have been calculated (Table 8.10).

Lubowe (1964), after analysing the stream junction angles of four regions of the Applachian plateau, came to the conclusion that 'the stream junction angle increases as the order of receiving streams increases'.

It is apparent from Table 8.10 that the observations of Lubowe are applicable in the cases of the mean stream junction angles of 15 out of the 30 selected drainage basins (Tentarigara,

TABLE 8.10 : **Mean stream junction angle (degree)**

Sl. No.	Basins	ORDER (U)					Mean for all orders
		1st	2nd	3rd	4th	5th	$\overline{(X)}$
1	Tentarigara	45.3	47.4	52.2	52.2	54.6	50.3
2	Koina	48.4	50.7	62.8	70.8	61.6	58.8
3	Kumirta	41.9	44.4	30.6	44.8	40.5	40.4
4	Sangjata	38.9	54.7	60.5	68.7	70.4	58.6
5	Sapu	50.4	54.8	65.7	66.4	81.6	63.8
6	Mahadeosalgara	52.3	52.1	40.2	61.4	70.4	55.3
7	Dadkagara	65.4	68.3	81.4	81.4	87.5	76.8
8	Kumbigara	55.4	50.4	27.3	37.3	45.4	43.16
9	Bodikuchagara	50.3	61.6	71.6	75.4	83.2	68.4
10	Roro	52.3	64.6	73.3	80.3	85.6	71.2
11	Ghagara	48.6	49.5	54.7	54.4	33.8	48.2
12	Bamni	44.3	43.2	40.4	48.9	50.3	45.4
13	Konkoa	30.2	32.4	39.9	42.4	50.6	39.1
14	Kongera	28.6	38.7	39.5	44.7	57.3	41.7
15	Torlo	35.4	39.6	54.4	44.8	60.5	46.9
16	Suru	49.7	67.6	55.5	52.6	70.4	59.2
17	Sankh	40.4	48.7	49.9	58.6	61.7	51.8
18	Kharsoti	38.9	44.8	52.6	61.7	69.7	53.5
19	Sapghara	46.7	59.7	44.6	64.6	57.7	54.7
20	Banki	64.6	70.5	71.4	77.0	—	70.9
21	Karanjia	31.4	35.6	38.9	44.0	—	37.5
22	Gara III	37.4	47.8	38.9	49.5	—	43.4
23	Toyargara	40.3	44.6	46.5	66.0	—	49.3
24	Sasagara	33.7	62.2	27.5	30.2	—	38.4
25	Kuchal	37.6	44.8	24.4	61.2	—	42.0
26	Kantajuria	44.2	48.4	44.8	62.2	—	49.9
27	Sona	28.5	36.6	45.4	49.6	—	40.0
28	Bhangbanga	27.7	37.4	40:8	41.9	—	36.9
29	Bagalduba	40.6	25.7	49.9	61.3	—	44.4
30	Dudh	44.5	48.6	49.4	37.4	—	45.0

Sangjata, Sapu, Dadkagara, Bodikuchagara, Roro, Konkoa, Kongera, Sankh, Kharsoti, Banki, Karanjia, Toyargara, Sona and Bhangbanga basins) which show continuous increase in mean stream junction angles with increasing orders. The remaining basins show some departures from the above mentioned rule though the general trend of entrance angles of these basins exhibits increase of angle with increasing orders. The basins, which do not follow the general rule of increase of entrance angles with increasing orders, are either elongated basins (Konkoa, Mahadeosalgara, Gara III, Torlo and Dudh) or those basins which drain over two different physiographic regions (after rising in the highlands flow over peneplain surfaces) wherein the lower order tributaries lying over the highlands make high entrance angles and, hence, the above rule is not followed by Suru, Kuchai, Bamni and Sasagara basins.

Schumm (1956) has pointed out that 'where the valley-side slope is steep in relation to the gradient of the master stream, a tributary tends to join at almost right angles; but where the gradient of master stream and valley-side slope are almost the same, the tributary virtually parallels the main channel, joining it at a small angle'. Thus, the entrance angles may be correlated with the stage of the basin development because 'the entrance angles in mature basins are significantly less than those in youthful basins' (Schumm, 1956). The average entrance angles of all the orders of the basins of the present study directly follow the observations of Schumm because the basins of the western highlands like Tentarigara (50.3°), Koina (58.8°), Sangjata (58.6°), Sapu (63.8°), Mahadeosalgara (55.3°), Dadkagara (76.8°), Kumbigara (43.16°), Roro (71.2°), Bodikuchagara (68.4°), Ghagara (48.2°), Bamni (45.4°), and Banki (70.9°), which are in late youth to early mature stages of basin development, have comparatively higher mean entrance angles than the basins of the central peneplain plateau (Bhangbanga = 36.9°, Karanjia = 37.5°, Sasagara = 38.4°, Konkoa = 39.1°, Sona = 40.0°, Kongera=41.7° Kuchai=42.0°, Gara III=43.4° and Torlo=46.9° basins), A marked difference in the average entrance angles is observed in the basins of the western and eastern Dalma highlands.

Geometry of shape of closed link

The geometry of basin shape reflects the genesis of the basins where low, medium and high circularity indices are related to youth, mature and peneplain stages of basin development respectively. The ideal shape of the basin is regarded as pear shape which is frequently disturbed by geological, relief and lithological factors. The shape of the closed link of the basin is determined by drainage area, perimeter and length of the master stream and it presents a good deal of fluctuations due to its dependence on aforesaid factors. The Horton's Form Factor* (1932), Miller's Circularity Index ** (1953), Schumm's Elongation Ratio *** (1966), and Chorley, Malm and Pogorzelski's Lemniscate † (1957) have been used to calculate the geometric shapes of 30 basins.

Kuchai basin has the most elongated shape as is indicated by 'F' (0.19), 'C' (0.36), and 'K' (1.35), values whereas Bhangbanga basin has the circular shape by having higher values of 'F' (0.53), 'C' (0.60) and 'R' (0.82) and lowest value of 'K' (0.47). On the basis of 'F' values, Bhangbanga (0.53), Sapghara (0.52), Suru (0.52) and Sankh (0.49) basins have nearly circular shapes whereas Kuchai (0.19), Kantajuria (0.25),

* Horton's Form Factor (F)

$$F=\frac{A}{L^2}$$

Where A=basin area and L=basin length

** Miller's Circularity Index (C)

$$C=\frac{4\pi A}{P^2}$$

Where P=basin perimeter

*** Schumm's Elongation ratio (R)

$$R=\frac{\text{diameter of the circle with same area as the basin}}{\text{basin length}}$$

Thus,

$$R=\frac{2\sqrt{\frac{A}{\pi}}}{L}=\frac{2}{\sqrt{\pi}}\sqrt{\frac{A}{L^2}}=\frac{2}{\sqrt{\pi}}\sqrt{F}$$

or

$$(F=\frac{\pi}{4}R^2)$$

† Chorley, Malm and Pogorzelski's Lemniscate (K)

$$K=\frac{L^2}{4A}$$

Note— (i) The values of 'F' vary from 0 (highly elongated shape) to the unity (circular shape).

(ii) The values of 'C' vary from 0 (a line) to unity (a circle).

(iii) The values of 'R' vary from 0 (highly elongated) to 1 (a circle).

(iv) The higher values of 'K' indicate elongated shape while lower values show circular shape.

Karanjia (0.25), Gara III (0.25), Sasagara (0.23), Bamni (0.26), Sangjata (0.26), Koina (0.29), Dudh (0.28) and Tentarigara (0.29) basins have elongated shapes because the 'F' values of these basins are below 30% (0.30). The remaining basins have 'F' values between 30% to 45% (0.30—0.45) which indicate semi-idented shapes.

It is apparent from Table 8.11 that the 'C' values are generally proportional to the values of 'F' with some exceptions.

TABLE 8.11 : **Shape indices**

Sl. No.	Basins	Circularity Index (C)	Elongation Ratio (R)	Form Factor (F)	Lemniscate (K)
1	Tentarigara	0.37	0.61	0.29	0.86
2	Koina	0.37	0.60	0.29	0.87
3	Kumirta	0.47	0.71	0.40	0.62
4	Sangjata	0.37	0.58	0.26	0.94
5	Sapu	0.43	0.74	0.43	0.58
6	Mahadeosalgara	0.45	0.68	0.36	0.69
7	Dadkagara	0.44	0·66	0.34	0.73
8	Kumbigara	0.63	0.63	0.31	0.81
9	Bodikuchagara	0.44	0.69	0.38	0.66
10	Roro	0.44	0.73	0.41	0.61
11	Ghagara	0.47	0.72	0.40	0.62
12	Bamni	0.31	0.57	0.26	0.97
13	Konkoa	0.27	0.62	0.30	0.83
14	Kongera	0.48	0.73	0.42	0.59
15	Torlo	0.47	0.65	0.33	0.75
16	Suru	0.51	0.82	0.52	0.48
17	Sankh	0.51	0.79	0.49	0.52
18	Kharsoti	0.45	0.67	0.35	0.71
19	Sapghara	0.49	0.83	0.52	0.46
20	Banki	0.43	0.63	0.31	0.80
21	Karanjia	0.42	0.56	0.25	1.01
22	Gara III	0.45	0.57	0.25	0.99
23	Toyargara	0.43	0.65	0.34	0.75
24	Sasagara	0.41	0.54	0.23	1.11
25	Kuchai	0.36	0.49	0.19	1.35
26	Kantajuria	0.42	0.56	0.25	1.02
27	Sona	0.57	0.64	0.42	0.78
28	Bhangbanga	0.60	0.82	0.53	0.47
29	Bagalduba	0.51	0.67	0.35	0.72
30	Dudh	0.44	0.60	0.28	0.88

Kumbigara has the highest 'C' value (0.63) indicating near circular shape but the 'F' value (0.31) puts it into the category of elongated shape which appears to be more correct when its perimeter is taken into account. On the basis of circularity index ('C'), Bhangbanga (0.60), Bagalduba (0.51), Sona (0.57), Sankh (0.51) and Suru (0.51) basins fall in the category of near circular shape as the 'C' values of these basins are above 50 per cent (0.50) whereas Kuchai (0.36), Koina (0.37) and Sangjata (0.37) basins have elongated shapes by having 'C' values below 40 per cent (0.40) while the remaining basins have 'C' values between 40 to 50 per cent.

The elongation ratio of Schumm (R) places Bhangbanga (0.82), Sapghara (0.83), Suru (0.80) and Sankh (0.79) basins in the categories of circular shapes because their 'R' values are above 75% (0.75) whereas the elongated shape is assumed by Kuchai (0.49), Kantajuria (0.56), Sasagara (0.54), Gara III (0.57), Karanjia (0.56), Bamni (0.57) and Sangjata (0.58) basins which have 'R' values below 60% (0.60). On the basis of 'K', the elongated shape is assumed by Kuchai (1.35), Sasagara (1.11), Kantajuria (1.02), Karanjia (1.01), Gara III (0.99), Bamni (0.97) and Sangjata (0.94) basins because 'K' value is above 0.90 whereas Bhangbanga (0.47) Sapghara (0.46), Suru (0.48), Sapu (0.58), Kongera (0.59) and Sankh (0.52) basins have circular shape by having 'K' values below 0.60.

It is apparent from the above discussion of four shape indices that Bhangbanga, Suru, Sankh and Sapghara basins are more circular than the other basins while Kuchai, Koina, Bamni, Tentarigara, Sasagara, Kantajuria and Sangjata basins have more elongated shapes. It is also apparent that the basins of the different physiographic regions exhibit mixed reactions to the geological structure, absolute relief, relative relief and slopes. The shape of the rivers of the western highlands has been guided by relief and structure. The parallel hill ranges have forced the rivers viz. Tentarigara, Koina, Banki, Sangjata, Kuchai and Bamni to assume elongated shapes whereas Sapu, Kumirta and Ghagara have developed their basins on weak Dharwarian structures which have favoured much branching of the streams and proportionate increase in basin area, basin perimeter and channel length due to back-wearing of watersheds because of lateral planation resulting into circular shapes. The basins of the central plateau have assumed near circular shapes due to the peneplain stage of the surface but the elongated shapes of Konkoa, Karanjia, Gara III and Sasagara basins are the results of dolerite dykes which have restricted these rivers to develop circular shapes. Suru, Sapghara and Kharsoti rivers of the Dalma highlands have developed circular shapes because these rivers have developed their basins over phyllite structures which have favoured uniform destruction of watersheds but parallel ranges and hills have prevented Bagalduba and Kantajuria basins to develop circular shapes.

EPILOGUE

The network analysis of 30 drainage basins of S.E. Chota Nagpur region reveals the fact that the branching of streams has been largely controlled by geological structure, lithological characteristics, general reliefs, plant cover, pedological characteristics, slopes and rainfall intensity. The basins having homogeneity in the above factors closely correspond to the different laws of drainage network but marked variation in the above factors has upset the normal functioning of drainage system. It is, therefore, dangerous to relate the morphometric properties of drainage network to the stages of basin development in Davisian style. The different mathematical models of Horton and Strahler related to the drainage network may not be applied in totality but linear relationship is very much validated.

SUGGESTED READINGS

Brice, J.C., 1964 : Channel patterns and terraces of the Loup Rivers in Nebraska, *U.S. Geological Survey*, *Proffesional Paper*, 422-D.

Chorley, R.J., 1969 : Drainage basin as the fundamental geomorphic unit, in *Water, Earth and Man*, edited by R.J. Chorley, Methuen and Co. Ltd., pp. 77-99.

Chorley, R.J., Malm, D.E.G. and Pogorzelski, H.A., 1957 : A new standard for estimating basin shape, *American Journal of Science*, Vol. 255, pp. 138-41.

Coates, D.R., 1958 : Quantitative geomorphology of small drainage basins of Southern Indiana, Office of Naval Research, Project NR 389-042, *Technical Report*, 10.

Giusti, E.V. and Schneider, W.J., 1965 : The distribution of branches in river networks, *U.S. Geological Survey*, Professional Paper, 442-G.

Horton, R.E., 1932 : Drainage basin characteristics, *Transaction of the American Geophysical Union*, Vol. 13, pp. 350-361.

Horton, R.E., 1945 : Erosional development of streams and their drainage basins : Hydrophysical approach to quantitative morphology, *Bull. Geol. Soc. Amer.* Vol. 56, pp. 275-370.

Kirkby, M.J., 1969 : Infiltration, through flow and overland flow, in *Water, Earth and Man* edited by R.J. Chorley, Methuen and Co. Ltd., pp. 215-238.

Leopold, L.B., Walman, M.G. and Miller, J.P., 1964 : *Fluvial Processes in Geomorphology*, Eurasia Publishing House (Pvt.) Ltd., New Delhi, p. 131.

Lubowe, J.K., 1964 : Stream junction angles in the drainage pattern, *American Journal of Science*, Vol. 262, pp. 325-39.

Miller, V.C., 1953 : A quantitative geomorphic study of drainage basin characteristics in the Clinch Mountain Area, Virginia and Tennessee, *Technical Report* No. 3.

Muller, J.E., 1968 : An introduction to hydraulic and topographic sinuosity index, *Ann. Assoc. Amer. Geogr.*, Vol. 58, No. 2, pp. 371-385.

Schumm, S.A., 1956 : The evolution of drainage systems and slopes in badlands at Perth Amboy, New Jersey, *Bull. Geol. Soc. Amer.*, Vol. 67, pp. 597-646.

Singh, Savindra, 1978 : *A geomorphological study of small drainage basins of the Ranchi plateau*, Unpublished D. Phil. thesis, Allahabad University.

Singh, Savindra and Srivastava, R., 1974 : A morphometric study of the tributary basins of upper reaches of the Belan river, *National Geographer*, Vol. 9, pp. 31-44.

Singh, Savindra and Srivastava, R., 1976 : Morphometric properties of the drainage-net developed over weathered upper Vindhyan bedrock : A case study of the lower reaches of the Belan basin, *Uttar Bharat Bhoogol Patrika*, Vol. 12, No. 1, pp. 49-59.

Strahler, A.N., 1952 : Dynamic basis of geomorphology, *Bull. Geol. Soc. Amer.*, Vol. 63, pp. 929-38.

Strahler, A.N. 1953 : Revisions of Horton's quantitative factors in erosional terrain (Abst), *Transactions of the American Geophysical Union*, p. 34.

Strahler, A.N., 1957 : Quantitative analysis of watershed geomorphology, *Transactions of American Geophysical Union*, Vol. 38, pp. 913-920.

Strahler, A.N., 1964 : Quantitative geomorphology of drainage basins and channel networks, in V.T. Chow's *Handbook of Applied Hydrology*, McGraw-Hill Book Co., pp. IV-39-76

Strahler, A.N., 1969 : *Physical Geography*, John Wiley and Sons Inc., New York, p. 483.

Tricart, J. and Cailleux, A., 1972 : *Introduction to Climatic Geomorphology*, Longman, London, p. 274.

Woldenberg, M.J., 1966 : Horton's laws justified in terms of allometric growth and steady state in open systems. *Bull. Geol. Soc. Amer.*, Vol. 77, pp. 431-34.

NINE

G. Padmaja and *Soundarvalli*

THE BEHAVIOURAL PATTERN OF BIFURCATION RATIOS IN GODAVARI AND KRISHNA BASINS (A.P.)

THE bifurcation ratio is expressed as the ratio of the number of streams, of an order to the next higher order, clearly speaks of an hierarchial system of a drainage basin. This ratio is of fundamental importance in drainage basin analysis as it is the foremost parameter to link with hydrological regime of a watershed under uniform lithological and climatic conditions.

According to Horton the bifurcation ratio as designated by Rb varies from a minimum of 2 in "flat or rolling drainage basins" to 3 or 4 in "mountainous or highly dissected drainage basins". Keeping this initial law of stream numbers in mind an attempt has been made here to analyse the nature of bifurcation ratios in two major river basins of Peninsular India, namely Krishna and Godavari.

These two rivers drain more than half of the drainage area of India traversing through varied geological formations ranging from Cuddapah system in the west to recent alluvium in the east. These two basins have been particularly chosen for the bifurcation analysis as they provide higher order tributaries and also the variations within the values is more due to not only geological formations but also due to varying climatic conditions in E-W direction.

As these two drainage basins are too large to compare, two sample basins from each of the two basins have been chosen

This Paper has been presented before II Annual Conference of National Association of Geographers at Tirupati (AP).

which are of sixth order (Appendix 1) for comparative analysis of bifurcation ratio, these are—

(i) Godavari catchment lying within topographical sheet nos. of 46 H and L.
(ii) Kadra basin also of the same toposheet belonging to major drainage system of Godavari basin.
(iii) Aleru basin restricted to 56 K toposheet.
(iv) Hagari river basin restricted to . . .

(A detailed index of the basins pertaining to the toposheet is provided in Appendix 1).

Analysis of bifurcation ratio

For these four basins of sixth order which are mainly chosen for a comparative analysis of Rb the values are derived and plotted on a graph Fig 9.1 where it can be noted that the plots are highly variable. The average bifurcation ratio (Rb) around 3.5 except for Hagari Basin speaks of otherwise high stability

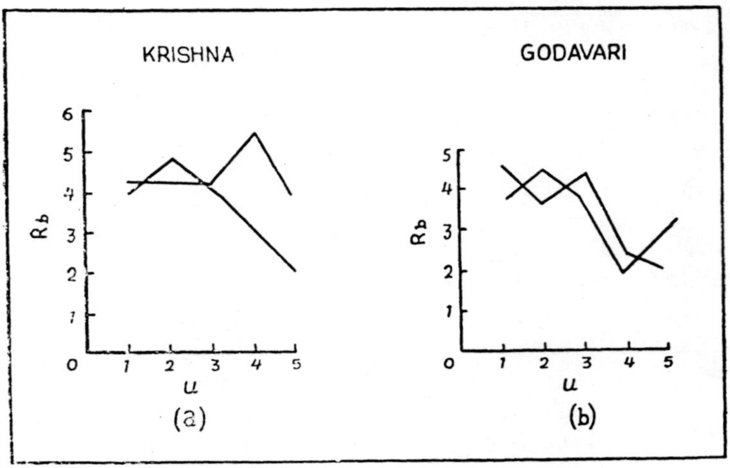

FIG. 9.1 : *A comparative analysis of Rb*

and small range of variation from region to region. But the variable nature of graph of order against bifurcation ratio (Rb) needs to be analysed and clarified. For this purpose, mean bifurcation ratios (Rb) and their standard deviations were computed for twenty fifth or fourth order basins of Krishna and Godavari system out of which there is an equal distribution of basins falling within each major drainage system.

The mean bifurcation ratios for individual orders range around 4 and 5 and give an impression of rugged and dissected terrain, Table 9.1.

TABLE 9.1 : **Mean Rb increases with decreasing order**

Krishna Basin	$\frac{Ns-1}{Ns-8}$	$\frac{Ns-2}{Ns-1}$	$\frac{Ns-3}{Ns-2}$	$\frac{Ns-4}{Ns-3}$	
Mean Rb	2.33	3.66	5.09	4.57	
Standard deviation	2.00	3.9	5.87	1.48	S=order of the main stream
Godavari Basin	$\frac{Ns-1}{Ns}$	$\frac{Ns-2}{Ns-1}$	$\frac{Ns-3}{Ns-2}$	$\frac{Ns-4}{Ns-3}$	
Mean Rb	2.67	3.88	4.88	4.37	
Standard deviation	4.14	5.92	3.73	2.38	

From this Table 9.1 it is clear that mean Rb increases with decreasing order upholding Horton's law of stream numbers. Rbs computed for lower order tributaries or Ns-1 are smaller than Rbs computed from lower order tributaries. The decreasing trend is interrupted in the case of $\frac{Ns\text{-}3}{Ns\text{-}2}$ rise or Rb represented as the ratio between second and third order rises to 5.09 and 4.88 respectively. The similarity in both the cases of Krishna and Godavari drainage systems speaks of similarity in drainage characteristics of both the basins.

The standard deviation values though are inclined towards lower values for higher order streams but are not consistent. The high value of 5.87 for second order to third order is highly inconsistent. But this value has a close correlation with high mean Rb for the same orders in Krishna Basin. In Godavari system the high value of 5.92 occurs for next successive rise in order in contrast to Krishna. But on the whole the larger deviation from the mean speaks of high variability in the nature of bifurcation ratios in both the drainage systems.

The next relation between area and Rb has been depicted by Fig. 9.2a and 9.2b. This figure brings out clearly a relation which can be stated as "basins of equal order, but variable areas tend to have the smallest areas, the ratios increase with increasing areas up to a certain size, beyond which the Rbs tend to become constant". This statement holds good parity for both the basins as is evidenced from the Fig. 9.2. As far as the Krishna Basin is concerned there is an initial drop in the area as figured by Peddavarka Basin, but rises again to become constant till a total distance of 1150 sq. kms., but Sindinur-

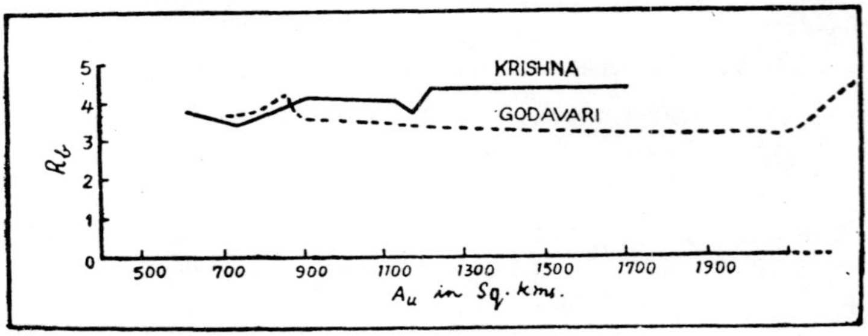

FIG. 9.2 : *Relation between Au and Rb*

nadi Basin marks a break as the line falls down depicting a drop in Rb but once again attains an increasing area with increasing Rb behaviour, and then remains constant throughout. So, it can be broadly divided into two sectors of identical behaviour. Though in segments but the statement comes to hold good in this basin. Godavari Basin too has a behaviour approximating the statement but in a different line. Here Kaunon and Muhr Basins with very high Rbs form the peak of the curves, but otherwise the rest of the curve follows a constant behaviour. The reason for the decreasing bifurcation ratios for increasing orders is basically due to branching process. It is apparent from the Table 9.2 (depicting the number of streams coalescing into next successive higher orders) that as the stream order increases the percentage of streams that coalesce into a higher order tributary decreases. This decrease is due to the decreasing area available. Thus, bifurcation ratios tend to become constant.

Distribution of tributaries within a drainage system

Melton (1958) defines a conservative drainage system as "one having the minimum number of channel segments necessary for the highest order of the system". With reference to this statement, the number of streams have been counted with a certain order lags, namely with two order lags (such as first order branches coalescing directly into third order, or 2nd order branches into 4th order) into one set of distribution similarly 3 order lags were also grouped into another set of distribution. It has been noted that the number of streams coalescing into higher order decreases with increasing order lag, with an exception of four order lag. In more than four cases it increases irrespective of the lag. Giusti and Schneider in their work on river networks and the distribution of branches within them give a logarithmic expression for the distribution of channels, from which they state that "more than one out of three streams similar to those studied here will have two major tributaries and one out of five, three major tributaries. A second order stream will most frequently have two first order streams as branches, a third order stream, two second order branches and so on". This is what Melton defined as conservative drainage system.

In this case, except for a few cases, the coalescing stream into next successive higher order is more than the percentage of streams draining into two or three order lags. These few exceptions are attributed to structural variations associated with climatic variations of the basins.

It has been noted that only first order streams coalesce into high order lag streams in contrast to next successive orders.

A. KRISHNA RIVER BASIN

1. *Hagari River* (U_6)

	U_1	U^2	U_3	U_4	U_5	U_6
U^2	1281	—	—	—	—	—
U_3	260	311	—	—	—	—
U_4	92	46	6.1	—	—	—
U_5	38	20	12	14	—	—
U_6	73	27	21	8	4	1
E	1744	404	94	22	4	1
Rb		4.3	4.3	4.3	5.5	4

Mean Rb=4.46

2. *Aleru River* (U_6)

	U_1	U_2	U_3	U_4	U_5	U_6
U_2	369	—	—	—	—	—
U_3	73	93	—	—	—	—
U_4	18	17	17	—	—	—
U_5	17	6	6	6	—	—
U_6	5	4	2	—	2	1
E	482	120	25	6	2	1
Rb		4.0	4.8	4.1	3	2

Mean Rb=3.58

3. *Wira River* (U_6)

	U_1	U_2	U_3	U_4	U_5	U_6
U_2	352	—	—	—	—	—
U_3	61	85	—	—	—	—
U_4	39	18	19	—	—	—
U_5	11	7	3	7	—	—
U_6	10	6	4	—	3	1
E	473	116	26	7	3	1
		4.08	4.46	3.7	2.3	3

Mean Rb=3.5

4. *Hindri River* (U_5)

	U_1	U_2	U_3	U_4	U_5	U_6
U_2	200	—	—	—	—	—
U_3	29	41	—	—	—	—
U_4	8	3	6	—	—	—
U_5	15	13	8	2	1	—
E	252	57	14	2	1	—
Rb		442	4.07	7.0	2	

Mean Rb=4.37

5. *Peddavarka River* (U_5)

	U_1	U_2	U_3	U_4	U_5
U_2	81	—	—	—	—
U_3	18	18	—	—	—
U_4	13	3	8	—	—
U_5	4	4	—	3	1
E	117	27	8	3	1
Rb		4.33	3.39	2.7	3.1

Mean Rb=3.36

6. *Sindinurnadi* (U_5)

	U_1	U_2	U_3	U_4	U_5
U_2	95	—	—	—	—
U_3	31	21	—	—	—
U_4	1	2	4	—	—
U^5	13	4	2	2	1
E	140	27	6	2	1
Rb		5.2	4.5	3	2

Mean Rb=3.7

7. *Eastern Tributary of Maneru.*

	U_1	U_2	U_3	U_4	U_5
U_2	130	—	—	—	—
U_3	19	20	—	—	—
U_4	15	16	8	—	—
U_5	3	6	—	3	1
E	167	42	8	3	1
Rb		4.0	5.3	2.7	3.0

Mean Rb=3.5

8. *Awarja* (U_5)

	U_1	U_2	U_3	U_4	U_5
U_2	173	—	—	—	—
U_3	50	35	—	—	—
U_4	15	4	8	—	—
U_5	28	11	5	4	1
E	266	50	13	4	1
Rb		5.3	3.0	3.2	4.0

Mean Rb=3.9

B. GODAVARI RIVER BASIN

1. 46 Godavari catchment

	U_1	U_2	U_3	U_4	U_5	U_6
U_2	243	—	—	—	—	—
U_3	68	52	—	—	—	—
U_4	23	11	16	—	—	—
U_5	19	11	5	5	–	—
U_6	9	4	—	—	2	1
E	362	78	21	4	2	1
Rb	4.6	3.7	4.2	2.5	2	

Mean Rb of the Basin=3.4

2. Kadva Basin

	U_1	U_2	U_3	U_4	U_5	U_6
U_2	255	—	—	—	—	—
U_3	54	71	—	—	—	—
U_4	52	15	20	—	—	—
U_5	4	12	2	6	—	—
U_6	7	2	1	—	3	1
E	373	100	23	6	3	1
Rb	3.7	4.4	3.8	2	3	

Mean=3.4

3. Suidphar

	U_1	U_2	U_3	U_4	U_5	U_6
U_2	605	—	—	—	—	—
U_3	180	129	—	—	—	—
U_4	73	20	27	—	—	—
U_5	11	15	9	8	—	—
U_6	13	12	6	3	3	1
E	882	176	32	11	3	1
Rb	5.0	5.5	2.9	3.7	3	

Mean for the basin=5.02

4. Dharna

	U_1	U_2	U_3	U_4	U_5
U_2	113	—	—	—	—
U_3	18	21	—	—	—
U_4	10	7	6	—	—
U_5	10	8	2	3	1
E	151	36	8	3	1
Rb	4.2	4.5	2.6	3	

Mean=3.6

5. Kaplan

	U_1	U_2	U_3	U_4	U_5
U_2	130	—	—	—	—
U_3	17	28	—	—	—
U_4	13	3	8	—	—
U_5	21	15	4	4	1
E	181	46	12	4	1
Rb	3.9	3.8	3.0	4	

Mean=3.68

6. Kaunn

	U_1	U_2	U_3	U_4	U_5
U_2	107	—	—	—	—
U_3	7	25	—	—	—
U_4	6	6	6	—	—
U_5	1	4	3	2	1
E	221	35	9	2	1
Rb	6.3	3.9	4.5	2	

Mean=4.2

7. Mulr River

	U_1	U_2	U_3	U_4	U_5
U_2	450	—	—	—	—
U_3	47	80	—	—	—
U_4	90	62	25	—	—
U_5	3	8	3	6	1
E	590	150	28	6	1
Rb	3.9	5.4	4.7	6	

Mean=5

8. Sikkunaru river

	U_1	U_2	U_3	U_4	U_5
U_2	75	—	—	—	—
U_3	25	16	—	—	—
U_4	10	5	5	—	—
U_5	3	2	—	2	1
E	113	33	5	2	1
Rb	3.4	6.6	2.5	2	

Mean of the basin=3.5

9. Satnadi

	U_1	U_2	U_3	U_4	U_5
U_2	78	—	—	—	—
U_3	14	22	—	—	—
U_4	6	2	8	—	—
U_5	4	1	—	2	1
E	102	25	8	2	1
Rb	4.1	3.1	4	2	

Mean of the basin = 3.3

10. Karanda

	U_1	U_2	U_3	U_4	U_5
U_2	112	—	—	—	—
U_3	9	22	—	—	—
U_4	37	13	10	—	—
U_5	8	3	—	4	1
E	166	38	10	4	1
Rb	4.4	3.8	2.5	4	

Mean of the basin = 3.7

11. Karanja (En)

	U_1	U_2	U_3	U_4	U_5
U_2	163	—	—	—	—
U_3	27	33	—	—	—
U_4	21	17	9	1	—
E	211	50	9	1	—
Rb	4.2	5.5	9.0		

Mean = 6.2

12. Swarna

	U_1	U_2	U_3	U_4
U_2	64	—	—	—
U_3	9	10	—	—
U_4	13	10	3	1
E	85	20	3	1
Rb	4.3	6.7	3.0	

Mean = 4.7

U = order of the stream.

APPENDIX I

KRISHNA RIVER SYSTEM

Sample No.	Sheet No.	Basin	Order	Near Rb	Area in sq. km.
1.	—	Hagari	U_6	4 46	—
2.	55 K	Aleru	U_6	3.58	—
3.	—	Wira	U_6	3.5	—
4.	57 I	Hindri	U_5	4.37	715.43
5.	57 I	Peddavarka	U_5	3.36	767.25
6.	57 A	Sindinurnadi	U_5	3.7	175.62
7.	65 C	En. Tributary of Maneru	U_5	3.75	609.06
8.	56 C	Awarja	U_5	3.9	148.25
9.	56 K	Chinnamusi	U_5	4.3	1211.56
10.	56 K	Shanirpet Nadi	U_5	4.05	900.37
11.	57 AVI	Qarchivanba	U_4	5.0	840.81
12.	56 K	Paler	U_4	5.8	650.37

GODAVARI RIVER SYSTEM

Sample No.	Sheet No.	Basin	Order	Near Rb	Area in sq. km.
1.	46 H & 46 Z	Godavari Catchment	U_6	3.4	2524.62
2.	46 H & 46 Z	Kadva	U_6	3.4	1852.50
3.	47 M & 56 A	Suidphar	U_6	5.02	3796.50
4.	47 P & 55 D	Dharna	U_5	3.6	710.50
5.	56 A	Kaplan	U_5	3.68	891.56
6.	47 M	Kaun	U_5	4.2	866.56
7.	47 I	Mulrnadi	U_5	5.0	6363.12
8.	56 I	Sikkumnau	U_5	3.5	921.06
9.	56 I	Sat nadi	U_5	3.3	1375.56
10.	56 G	Western Karanja	U_5	3.7	776.81
11.	56 H	Eastern Karanjia	U_4	6.2	1173.87
12.	56 I	Swarna	U_4	4.7	741.62

INDEX